T0189519

# Advances in Intelligent Systems and Computing

Volume 525

**Series editor**

Janusz Kacprzyk, Polish Academy of Sciences, Warsaw, Poland
e-mail: kacprzyk@ibspan.waw.pl

## About this Series

The series "Advances in Intelligent Systems and Computing" contains publications on theory, applications, and design methods of Intelligent Systems and Intelligent Computing. Virtually all disciplines such as engineering, natural sciences, computer and information science, ICT, economics, business, e-commerce, environment, healthcare, life science are covered. The list of topics spans all the areas of modern intelligent systems and computing.

The publications within "Advances in Intelligent Systems and Computing" are primarily textbooks and proceedings of important conferences, symposia and congresses. They cover significant recent developments in the field, both of a foundational and applicable character. An important characteristic feature of the series is the short publication time and world-wide distribution. This permits a rapid and broad dissemination of research results.

## Advisory Board

More information about this series at http://www.springer.com/series/11156

Ryszard S. Choraś

Editor

# Image Processing and Communications Challenges 8

8th International Conference, IP&C 2016
Bydgoszcz, Poland, September 2016
Proceedings

 Springer

*Editor*
Ryszard S. Choraś
Institute of Telecommunications
  and Computer Sciences
UTP University of Science and Technology
Bydgoszcz
Poland

ISSN 2194-5357          ISSN 2194-5365   (electronic)
Advances in Intelligent Systems and Computing
ISBN 978-3-319-47273-7          ISBN 978-3-319-47274-4   (eBook)
DOI 10.1007/978-3-319-47274-4

Library of Congress Control Number: 2016952533

Printed on acid-free paper

This Springer imprint is published by Springer Nature
The registered company is Springer International Publishing AG
The registered company address is: Gewerbestrasse 11, 6330 Cham, Switzerland

# Preface

We welcome you to the International Conference on Image Processing and Communications, IP&C 2016. The present volume contains the proceedings of the International Conference on Image Processing and Communications, IP&C 2016, held at Bydgoszcz, Poland, September 7–9, 2016.

IP&C 2016 was organized by the UTP University of Technology and Sciences and was hosted by the Institute of Telecommunications and Computer Sciences of the UTP University.

The IP&C 2016 brought together the researchers, developers, practitioners, and educators in the field of image processing and computer networks. IP&C has been a major forum for scholars and practitioners on the latest challenges and developments in IP&C.

The conference proceedings contain 37 papers which were selected through a strict review process, with an acceptance rate at 57 %. In all, 37 papers entered the review process and each was reviewed by two independent reviewers using the double-blind review method. There were also two invited talks by Massimo Ficco and by Damian Karwowski.

The presented papers cover all aspects of image processing (from topics concerning low-level to high-level image processing) and modern communications.

The organization of such an event is not possible without the effort and the enthusiasm of the people involved. The success of the conference would not be possible without the hard work of the local Organizing Committee.

We would like to thank all authors for the effort they put into their submissions.

Last but not least, we are grateful to Springer for publishing the IP&C 2016 proceedings in their Advances in Intelligent Systems and Computing series. Finally, we thank the Springer team for helping us in the final preparation of this AISC book.

I hope that all of the attendees found the conference informative and thought-provoking.

Ryszard S. Choraś
Conference Chair
IP&C 2016

# Organization

## Organization Committee

### Conference Chair

Ryszard Tadeusiewicz     Poland
Ryszard S. Choraś     Poland

## International Program Committee

Kevin W. Bowyer     USA
Dumitru Dan Burdescu     Romania
Christophe Charrier     France
Leszek Chmielewski     Poland
Michał Choraś     Poland
Andrzej Dąbrowski     Poland
Andrzej Dobrogowski     Poland
Marek Domański     Poland
Kalman Fazekas     Hungary
Ewa Grabska     Poland
Andrzej Kasiński     Poland
Andrzej Kasprzak     Poland
Marek Kurzyński     Poland
Witold Malina     Poland
Andrzej Materka     Poland
Wojciech Mokrzycki     Poland
Sławomir Nikiel     Poland
Zdzisław Papir     Poland
Jens M. Pedersen     Denmark

Jerzy Pejaś                 Poland
Leszek Rutkowski            Poland
Khalid Saeed                Poland
Abdel-Badeeh M. Salem       Egypt

# Organizing Committee

Sşawomir Bujnowski
Piotr Kiedrowski
Damian Ledziński
Zbigniew Lutowski
Adam Marchewka - Publication Chair
Beata Marciniak
Tomasz Marciniak
Ireneusz Olszewski
Karolina Skowron - Conference Secretary
Mścisław Śrutek
Łukasz Zabłudowski

# Contents

# Image Processing

# 20 Years of Progress in Video Compression – from MPEG-1 to MPEG-H HEVC. General View on the Path of Video Coding Development

Damian Karwowski[✉], Tomasz Grajek, Krzysztof Klimaszewski,
Olgierd Stankiewicz, Jakub Stankowski, and Krzysztof Wegner

Chair of Multimedia Telecommunications and Microelectronics,
Poznań University of Technology, Polanka 3 Street, 60-965 Poznań, Poland
dkarwow@multimedia.edu.pl

**Abstract.** Compression of moving images has opened unprecedented opportunities of transmission and storage of digital video. Extraordinary performance of today's video codecs is a result of tens of years of work on the development of methods of data encoding. This paper is an attempt to show this history of development. It highlights the history of individual algorithms of data encoding as well as the evolution of video compression technologies as a whole. With the development of successive technologies also functionalities of codecs were evolving, which make also the topic of the paper. The paper ends the attempt of authors' forecasting about the future evolution of video compression technologies.

## 1 Video Compression – What Is It About?

The video, as we know it, no longer is a set of celluloid tapes as it used to be in the past days. Currently, practically the only form of video is a piece of digital information. Digital video is a way of storing the information about the series of rectangular array of pixels frames, representing a light in 3 frequency sub-bands: red, green and blue, changing at least 25 times per second.

The problem is that the direct representation of digital video would require an enormous number of bits. For example in the case of the nowadays television, each frame has a resolution of $1920 \times 1080$ pixels for each of the three color components, which results in more than 6 MB (megabyte: $10^6$ bytes) of data. Taking into account that we have 25 such frames per second we would require 155 MB of space to store a single second of a video. And it is not the end, currently 4 K, 8 K or even higher resolution videos are considered. Therefore, transmission, and even storage of uncompressed video is not feasible.

In order to allow efficient video transmission, the digital signal representing the moving pictures must be compressed.

But how does one compress the video? Generally there are two approaches: to use some mathematical tricks to squeeze the data to the edge of the entropy of the data, or just to throw out less important data contained in the video.

© Springer International Publishing AG 2017
R.S. Choraś (ed.), *Image Processing and Communications Challenges 8*,
Advances in Intelligent Systems and Computing 525, DOI 10.1007/978-3-319-47274-4_1

The limits of the first approach are expressed by the entropy of the coded data. In the second approach, the question is which data are unimportant, and can be throw out. The desired way is to discard the information that is either not seen (perceived) by the viewers, or not important for the overall impression. This is the dominant approach nowadays, since it gives by the orders of magnitude better compression ratios than the lossless from definition entropy based coding.

## 2    What Algorithms and Compression Technologies Have Been Developed?

### 2.1    Algorithms of Data Encoding

In this section the main algorithms of data encoding will be recalled, which are well recognized in the field of image/video compression. From the point of view of contemporary image/video compression these algorithms should be considered as a single functional block in a codec, rather than a full compression technology.

**Entropy Coding.** The beginning of the development of data compression methods dates back to the late 40s of the last century, when Claude E. Shannon has presented the results of his work [24], which clearly showed what are the basic limitations of data compression, e.g. how much the data can be compressed without losing information. Video signal is not an exception here.

Shannon's work became the foundation for many methods of statistical coding that were developed later, which are known in the literature as entropy coding methods [23]. At least, the following methods should be mentioned here: Huffman coding [8] (year 1952), and its widely used special cases Golomb codes [7] (year 1966), exponential Golomb codes [28] (1978), LZW, arithmetic coding [1] (year 1963, but suggested earlier by Elias), and recently developed by Polish scientist ANS method [6] (year 2005).

From the beginning of development of the video compression, there were attempts to treat moving pictures as any other digital signal. Thus, one tried to encode it with the use of entropy coding only. However, it turned out very quickly that very little compression of video can be achieved in this way (fife-fold more or less), due to the value of entropy of the original video data.

**Prediction.** Taking into account the above, additional coding tools were strongly needed, that would be an essential complement to the methods of entropy coding. First ideas of such tools boiled down to prediction of image samples based on the already transmitted ones. The goal was to transmit only the difference between prediction signal and the actual one, instead of the original image pixel.

An example of such tools were developed in 1950 in the form of Differential Pulse Code Modulation (DPCM) technique [29]. More advanced approach that was developed later is INTRA directional prediction, that is commonly used in all contemporary video codecs (like AVC and HEVC).

Even with the use of advanced INTRA prediction the efficient encoding of a video was still a serious problem. In this context, a breakthrough was in 1981, when it was developed the technique of predictive coding of a video data with the motion estimation and compensation [18]. It is commonly known as INTER coding. With this method it was possible to predict accurately the motion in a video sequence over the time, which became the basis for a very efficient compression of a video.

**Transform Coding.** Discussed in the previous sub-section predictive coding leads to decorrelation of the data on whom this algorithm is realized. The same goal can be achieved by the use of Discrete Cosine Transformation (DCT) [2]. This is a type of a transform coding, and was developed in year 1974. In this transformation the input signal is represented with a cosinusoidal components, which in the case of image/video data can be a source of a significant reduction of a bitrate. Taking into consideration properties of human visual system it is worth to use this transformation in a combination with a lossy coding. Thus, transform coding of a video data followed by quantization of transform coefficients make an approach which is in a common use today.

## 2.2 Video Compression Technologies

In the previous section, examples of algorithms of data compression have been presented. Contemporary video codec must be treated as a collection of a number of such algorithms resulting in a technology (and not a single algorithm) of video compression.

**Hybrid Video Compression.** A joint application of the predictive coding (INTRA and INTER) together with the DCT-based lossy transform coding and the entropy coding of the data is commonly known in the literature as hybrid video compression. This technology of video encoding is in common use, and will be a topic of a more detailed considerations in the following part of the paper.

**Wavelet Image/Video Compression.** The hybrid video coding was not the only one, which in the 80s has given a justified hopes for efficient representation of the image, also moving images. Another such method was the wavelet coding (or subband coding) that uses Discrete Wavelet Transformation (DWT) [3,4,30]. This method was developed especially intensive since the 80s until 2004. The performance of solutions of the wavelet compression, that were developed for still images was so astonishingly high, when compared to other available then techniques (a reflection of this performance were the capabilities of the JPEG2000 encoder [12]), that many people believed that the wavelet compression will replace the hybrid codecs in a short time. Thereby, there was increasing the pressure to repeat the success of JPEG2000 also for the purpose of a video compression, which has motivated many laboratories to work on this compression technique. Undoubtedly, a breakthrough here was the beginning of the 90s, when it was proposed the concept of a three-dimensional video coding with motion compensation (e.g. works of Ohm from 1992 – 1994 [20–22]). The success of this method caused, that the hybrid coding techniques and the wavelet techniques became for yourself a direct competition.

Situation changed in 2004 when Moving Picture Experts Group (MPEG) and Video Coding Experts Group (VCEG) looked for the best technology for scalable video coding [16]. There were number of proposals based both on the hybrid technique and wavelet coding. But at that time hybrid approach outclassed the other proposals [15]. Hybrid compression proved to be the best for compressing a video.

**Parametric Video Compression.** Video compression methods that have been cited above have become the subject of numerous applications, both in the international compression standards [9–11,13,14,17], as well as commercial codecs [25–27,31]. Over the past 20 years, other methods have been also developed, but they have not found wide practical application so far. Among these methods a special attention should be payed on algorithms of parametric coding of an image texture [19]. In the case of a texture of a respectively high degree of complexity, the data describing the texture are not sent to the decoder in general. Instead of compressing or transmitting information about individual pixels, one can try to describe the video synthetically, in words, like in the following sentence: White house built from orange bricks. Due to the bitrate of data describing parameters of the texture is incomparably smaller than a data stream of traditional encoding, one can expect its practical use in the future.

## 3   Milestones in History of Hybrid Video Compression Development

The history of hybrid video compression dates back to 1989 when H.261 standard has been worked out by ITU-T. Although this standard was a set of very simple coding techniques (it was DCT-based lossy coding realized in a fixed-size blocks of the image), it gave a possibility to transmit a video over ISDN networks. At the same time (more or less) the works were continued on the MPEG-1 standard of ISO/IEC, whose purpose was to compress a raw digital video under the bitrate up to 1.5 Mbps, when achieving the VHS-quality video [14].

But the standard that really revolutionized the way of video transmission and storage was MPEG-2 [9]. Developed over 20 years ago standard of ISO/IEC has proven to be vastly popular, its popularity on the market was a sign of a great success of digital television. Not only was this the beginning of the process of replacing the older analog television, but it also introduced to our homes the theater systems with a video signal of unprecedented quality. In this way, the MPEG-2 became the first sign of a certain breakthrough.

Over the years, however, expectations of the users for even higher quality of encoded video (e.g. higher video resolution, less artifacts) as well as for the larger amount of available video content (e.g. higher number of digital channels in TV) were still growing. The MPEG-2 technique offered nearly 50-fold compression of a video (while ensuring high quality of images), and that proved to be insufficient. For this reason, the hybrid compression techniques were in the following years the subject of intensive works and improvements, that resulted in developing such standards as H.263, AVS, VC-1, H.264 (AVC) [21,22,25]. It was all the results of works that were carried out in years 1995 – 2003. Although all those standards were primarily aimed at different applications, each and every one of them introduced some new concepts, offered some improvements. Fundamental change was the introduction of a wider range of image block sizes, in which the encoder can perform data compression.

In the last 3 years, the works resulted in development of the newest, high-performance video compression technology, known as High Efficiency Video Coding (HEVC) [13]. Compared to older technologies of video compression, HEVC means more coding tools, and even higher adaptability of the size of image blocks, in which compression of data is carried out. With about 200-fold compression of a video (being a result of further improvements of the AVC technology), the new HEVC technology aims to meet the expectations of the contemporary market. Of course, this high efficiency comes at a price. The major problem is very high complexity of algorithms used in HEVC, so currently researchers look for the ways of reducing the complexity while maintaining high efficiency.

## 4   Is There Any Pattern in the Chaos?

As it can be seen, the variety of solutions for the problem of video compression is broad. But surprisingly, there can be seen a pattern in the development of the new standards. Similarly as for the globally acclaimed Moore's law for the development of semiconductor devices, there is a general rule for the efficiency of the consecutive generations of video coders, expressed by Domański's curve in [5]. Beginning from the mid-nineties of the 20th century, approximately each nine years, the progress in the domain of video coding is concluded by the development of a new standard. And each new standard is approximately twice as efficient as the previous one in encoding the contents contemporary to the given video compression standard. Of course, as for the Moore's law, also for the video compression we cannot expect the trend to continue indefinitely, but it seems that we are still at the stage of continuous progress of video compression methods.

## 5   New Replacing the Old

It needs to be noted, though, that the newer compression methods are much more complex and much more sophisticated than the previous ones, and require much more computational power. Additionally, the proliferation of the new standard is not immediate, since the industry needs some time to adapt to the new ways of processing of the digital video. Therefore, the actual progress, as seen in the

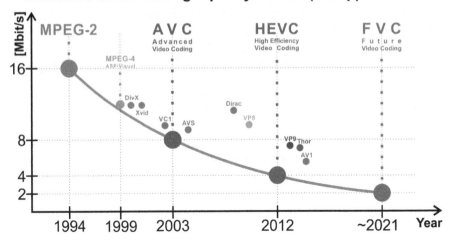

**Fig. 1.** Ilustration of performance of successive generations of video encoders

offerings of manufacturers, is not an abrupt one, but rather a steady increase of coding efficiency between the releases of the new standards (Fig. 1).

One of the most important factors to keep in mind when considering the upgrade to the new standard is the fact, that at the beginning there is no tools to support new codec, no good practices are established and there are lots of bugs, both in software and hardware. A certain time is needed for the new standard to settle down and become usable. Trying to implement the new standard too soon often means becoming a beta tester and pioneer in the unknown.

In this connection, compression efficiency of the first industry-oriented real time video encoders is not so high. It needs some time to develop the optimal way of control of a video encoder that enables full usage of potential of the technology. General dependence of the efficiency of the reference encoder and the product encoder has been illustrated in the figure below for MPEG-2, AVC and HEVC technologies.

## 6  What Was the Driving Force for Video Compression Enhancements?

The most important drive for the enhancements in the process of video coding was the rising expectations of the viewers regarding the quality of the video. Also the exponentially rising computational power of contemporary computers and – more generally – the hardware used for recording, processing and displaying the content contribute to the development of the new standards.

Also the progress of the science and development of the new methods and algorithms make a good foundation for development of new standards, since the

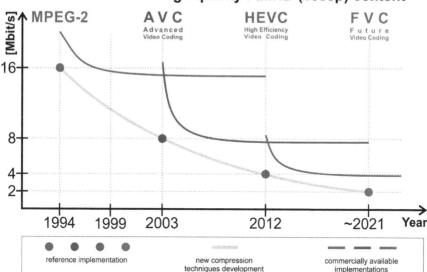

**Fig. 2.** Evolution of performance of commercially available encoders

companies that invent the new methods tend to capitalize on their effort by placing their developments into the coding standards (Fig. 2).

Also the changes in the way the data is used and consumed forces the development of the new standards, in order to better meet the requirements of certain applications and to provide the new functionalities required by the given applications.

## 7   Evolution of Functionalities

With the progress of development of compression methods, apart from efficiency, also the functionality started to matter. Not only was the efficiency started to play a role, but also other features that the codecs were expected to have. In order to maintain high coding efficiency while providing those new requested features, the coders needed to evolve by implementing a set of necessary functionalities.

One of the first significant features implemented in coders was the ability to efficiently encode interlaced sequences. In the advent of digital video, there was the necessity to compress also the video intended to use on television. In those days, television used interlaced video (each picture was divided into two fields – one contained even lines of image, while the other field contained the odd lines). Specialized prediction modes were introduced to help compress those interlaced video sequences efficiently, as well as different schemes of encoding the data. In the modern television, the necessity of using interlaced video is no longer valid (the display technology changed a lot since the days of the first video coders) and the recent standard does not support interlaced video coding.

Another important feature that the codec was expected to have is the error resilience. During transmission or storage, the bitstream (i.e. compressed video data) may become corrupted. Even the change of a single bit value may render the bitstream unusable, since the entropy decoders cannot handle any errors. To prevent that, several features were introduced in the codec that help to recover from such errors, at a cost of slightly lower compression ratio. The same mechanisms make it possible to decode the bitstream from the middle – such a situation is necessary for streaming of video and for digital television, since the users may start to watch the video at any time during transmission.

The next big feature introduced in video coders was so called scalability. This mechanism is intended to provide means to prepare a single bitstream that contains separate parts that make it possible to decode the video in different ways. There are different kinds of scalability:

- Time scalability – when only certain frames are decoded, and the data for the other frames can be omitted
- Spatial scalability – when only a lower resolution images can be decoded, while omitting data for higher resolution version of sequence
- Quality scalability – when the quality of the reconstructed video depends on whether the whole bitstream is decoded or only the part of bitstream is available for decoding.

The most obvious way to implement such features is to include several video streams in a single bitstream, but this leads to much larger file sizes (and thus, much wider bandwidth necessary to transmit it). But the scalability features are expected to share some parts of bitstream in order to prevent the redundancy. This means that there is a basic part of bitstream, necessary to decode any version of video, that provides the lowest quality, and there are some parts of the bitstream that improve the quality of the basic bitstream, by adding more information that enables to increase the resolution of the image, increase the quality of the images or increase the framerate. With the widespread online video streaming, this feature gains even more importance nowadays.

Another widely desired feature is the ability to compress stereoscopic video (video for left and right eye that provides the sense of depth of the scene) and even multiview images. Those functionalities were intended to be used to compress data for tv sets of all kinds that enable the 3D effect – be it shutter technology, polarizer technology, optical barrier or any other 3D display technology.

The idea of transmission of stereoscopic images was soon replaced by the multiview video, where there are much more than 2 views that are encoded simultaneously. All those techniques are exploiting the fact that most of the parts of the images from different views are similar, so they present an additional opportunity for prediction of the contents of the encoded image.

The fresh extension of this idea is to encode depth maps along with a few views in order to be able to synthesize all the necessary views after decoding the video and depth maps, presumably at the decoder size. This functionality is called 3D video.

Another feature that focuses more attention due to the development of better display technologies is so called High Dynamic Range (HDR) coding. This functionality allows to encode video with more than 8 bit data for luminance and chrominances. This is a natural follow up for the development of video coders after the provision to encode very high definition video. Ultra high definition displays are able to display very real-life images, but the video data provided must not only be of sufficiently high spatial resolution, but also must have high dynamic range, so that the wider range of values for each pixel can be used. In this context it is also interesting to mention the idea of moving from a standard RGB color space of displays, to a space that allows a more lifelike images by adding a fourth basic color – yellow. The development of RGBY displays has to be followed by the proper codec that could be used to provide the data for such a display.

As the cameras get cheaper and cheaper and provide even higher resolution all the time, it is also possible to imagine a system that captures the images of the whole perimeter at once – those are so called 360 degrees systems. This kind of data is useful in all kinds of virtual reality applications. Of course, there is a need for a codec that is able to compress such a kind of video data.

Another new feature that the coders developed in future are expected to have are the abilities to efficiently encode computer screen content. In the past years it would be considered wasteful to record the video of a screen content, since usually it would be much more efficient (in terms of data that need to be transmitted to the receiver or stored on the disk) to simply record all the actions performed on the screen and then replicate them at the decoder side by simply replicating them.

Compared to this method, the recording of the screen content video is a much more straightforward and much easier way of storing the actions performed on-screen. There are also less compatibility issues in case of simple recording of the screen content as a video.

One of the recently added functionality is the set of modifications called Green MPEG. Those features are there to save power required to process the data at the encoder, transmission and the decoder side. This way it is, for example, possible to prolong the battery life of a mobile device that decodes the video and thus, also make the whole process of decoding a less burdening to the environment. This can be obtained by sending the additional information to the decoder to turn off or adjust the operation frequency of certain modules that will not be necessary or be used less intensively for decoding of the video or a set of frames. It is also possible to adapt the displaying process to the properties of the display, in order not to waste the energy unnecessarily, for example for the backlight.

## 8  Developing New Encoders – Change of Paradigms

The earliest video compression standards (like. MPEG-2 Video) have been developed for usage in a dedicated encoding and decoding devices i.e. hardware encoders and decoders. The codec implementation cold be easily parallelized at high level (slice, frame), as well as at low level (block level) which is especially useful for hardware implementation. Briefly speaking, the goal was to develop the standard that is hardware-friendly.

Next generations of video compression technologies (H.263, AVC) have been developed in time when personal computers became powerful enough to do encoding and decoding of a video. Therefore, most compression tools that were developed in the context of H.263 and AVC took into account an efficient operation when realized in the software, and not necessary in the hardware. This led to evolution of a highly complex tools (like CABAC algorithm), which are extremely difficult to parallelize and not suitable for hardware implementations. Moreover, in above mentioned generation of compression standards there is a lack of effective parallelization techniques.

The significant turnaround took place during the development of HEVC. The coding tools complexity have been evaluated including both software and hardware implementations. The coding tools (like entropy encoding) have been substantially modified in order to allow parallel processing of data in the hardware. Moreover, a number of parallel processing possibilities have been introduced (like parallel merge, wavefront processing order, picture tiles, etc.) in order to speed up computations in the software edition of codecs.

Today we have a quite large number of different standards of video encoding. Individual standards are mutually incompatible (on the level of the syntax of encoded data stream), however most of the coding tools that are used in codecs (like motion estimation or DCT transformation) are the same. The fundamental difference is that the individual codecs may use these tools with a different input parameters, like the size of the image block in which motion estimation is carried out.

Thus, instead of implementing in a device each of the standard independently, it is better to implement a set of common functional blocks (like DCT or motion estimation), and lunch the algorithms with a given parameters, depending from the requirements of a given standard. Modern devices are built just according to this practice, which actually contain some fixed functional blocks and a general purpose core, which controls these blocks.

The architecture of devices mentioned above leads to the idea of Reconfigurable Video Coding (RVC) where encoded video bitstream contains the list and description of basic functional blocks which are necessary to decode the content.

## 9   What Will Come in Upcoming Years?

Of course it is impossible to predict accurately the future, nevertheless it is interesting to summarize the predicted directions in which the upcoming video compression technology will be developed. Basing on the current trends, it is anticipated that the video compression will be still constantly evolving, although without any dramatic leaps. Thanks to the progress in the area of computational hardware, e.g. processors, it will be possible to employ more and more advanced algorithms, using more and more memory, which altogether will yield higher and higher compression efficiency. As the golden fleece of computing is now usage of parallelism it is expected that the future video codecs will be designed in a way allowing for usage of multiple cores and processors with only a minor loss of efficiency.

An important factor to be considered is that the recent observations are that the pace of advancement in computational power has slowed and the progress anticipated with Moores law would soon reach saturation. Even if this will turn to be true, the speed of progress in computing should be sufficient to cater the needs of compression, because, at the same time, the image resolution enhancement is also slowing down. Even while currently 4 K, 8 K or even higher resolutions are considered, most of the content is displayed on 7 inch mobile devices on which Full-HD resolution is more than enough due to limitations of the humans eye. In fact, there is very little interest in representing the same field of view with higher resolutions. What is considered instead is an extension of the idea of narrow window-like vision to a wider case e.g. very wide panorama vision or even 360 vision which corresponds to representing pixels on the surface of a sphere. Such videos are already being used in applications related to light-field and virtual reality. In such, the viewers see only a part of the transmitted video, e.g. using portable viewing glasses like Oculus Rift. Therefore, partial decodability of the video stream may be of great need, because it would allow for reduction of computational costs and thus lower energy consumption. Energy consumption, connected with both battery life and power dissipation, is a fundamental concern of designers for mobile devices. The share of mobile displays in the video world is expected to rise even more and thus energy consumption might become a flagship of video compression developments. Initiatives like Green MPEG show that holistic approach to reduction of energy consumption, that will go beyond straight-forward implementation optimizations, e.g. developing energy-aware compression technologies, might be one of directions.

To summarize, it can be said the developments in video compression technology will be rather focused on new functionalities than bare compression efficiency improvements.

**Acknowledgment.** The research project was supported by The National Centre for Research and Development, Poland. Grant No. LIDER/023/541/L-4/12/NCBR (2013–2016).

# References

1. Abramson, N.: Information Theory and Coding. McGraw-Hill, New York (1963)
2. Ahmed, N., Natarajan, T., Rao, R.K.: Discrete cosine transform. IEEE Trans. Comput. **C−23**, 90–93 (1974)
3. Daubechies, I.: Orthonormal bases of compactly supported wavelets. Commun. Pure Appl. Math. **41**, 909–996 (1988)
4. DeVore, R., et al.: Image compression through wavelet transform coding. IEEE Trans. Inf. Theory **38**(2), 719–746 (1992)

5. Domański, M.: Approximate Video Bitrate Estimation for Television Services, ISO/IEC JTC1/SC29/WG11 MPEG 2015, M36571, Warsaw, Poland, 20–27 June 2015
6. Duda, J.: Asymmetric numeral systems: entropy coding combining speed of Hyffman coding with compression rate of arithmetic coding. arXiv: 1311.2540
7. Golomb, S.W.: Run-length encoding. IEEE Trans. Inf. Theory IT–12, 399–401 (1966)
8. Huffman, D.A.: A method for the construction of minimum-redundancy codes. In: Proceedings of the I. R. E, pp. 1098–1101, September 1952
9. ISO/IEC 13818-2 International Standard, "Generic Coding of Moving Pictures, Associated Audio", Part 2: Video, 2nd edn. (2000)
10. ISO/IEC 14496-10, International Standard, Generic Coding of Audio-Visual Objects, Part 10: Advanced Video Coding, 6th edn., 2010, take ITU-T Rec. H.264, Edition 5.0 (version 11) (2010)
11. ISO/IEC 15444-1, International Standard, JPEG: Core coding system, 2nd edn., 2004, also: ITU-T Rec. T.800, 2nd edn. (2002)
12. ISO/IEC 15444-1 and ITU-T Rec. T.800, Information Technology JPEG 2000 image coding system (2000)
13. ISO/IEC and ITU-T, High Efficiency Video Coding (HEVC), ISO/IEC 23008-2 (MPEG-H Part 2)/ITU-T Rec. H.265 (2013)
14. ISO/IEC IS 11172-2 International Standard, Coding of Moving Pictures, Associated Audio for Digital Storage Media at up to about 1.5 Mbps, Part 2: Video (1993)
15. ISO/IEC JTC1/SC29/WG11 MPEG2004/M110737, Subjective test results for the CfP on Scalable Video Coding Technology, Munich, March 2004
16. ISO/IEC JTC1/SC29/WG11, N6193, MPEG Call for proposals on scalable video coding technology
17. ITU-T Rec. H.263, Video Coding for Low Bitrate Communication (2005)
18. Jaswant, R., Anil, J., Jain, K.: Displacement measurement and its application in interframe image coding. IEEE Trans. Commun. - TCOM 29(12), 1799–1808 (1981)
19. Ndjiki-Nya, P., Stüber, C., Wiegand, T.: A new generic texture synthesis approach for enhanced H.264/MPEG4-AVC video coding. In: Atzori, L., Giusto, D.D., Leonardi, R., Pereira, F. (eds.) VLBV 2005. LNCS, vol. 3893, pp. 121–128. Springer, Heidelberg (2006). doi:10.1007/11738695_17
20. Ohm, J.-R.: Temporal domain subband video coding with motion compensation. In: Proceedings of the IEEE International Conference on Acoustics, Speech and Signal Processing, vol. 3, pp. 229–232 (1992)
21. Ohm, J.-R.: Three-dimensional motion-compensated subband coding. In: Proceedings of International Symposium on Video Communications and Fiber Optic Services, SPIE, vol. 1977, pp. 188–197 (1993)
22. Ohm, J.-R.: Three-dimensional subband coding with motion compensation. IEEE Trans. Image Process. 3, 559–571 (1994)
23. Salomon, D., Motta, G.: Handbook of Data Compression. Springer, London (2010)
24. Shannon, C.E.: A mathematical theory of communication. Bell Syst. Tech. J. 27, 379–423 (1948)
25. SMPTE Standard for Television: VC-1 Compressed Video Bitstream Format and Decoding Process, ANSI/SMPTE 421M (2006)
26. SMPTE Standard: VC-2 Video Compression, SMPTE 2042-1:2009 (2009)
27. SMPTE Standard: VC-3 Picture Compression and Data Stream Format, SMPTE2019-1:2005 (2005)

28. Teuhola, J.: A compression method for clustered bit-vectors. Inf. Process. Lett. **7**, 308–311 (1978)
29. U.S. patent 2605361, C. Chapin Cutler.: Differential Quantization of Communication Signals, filed June 29, 1950, issued 29 July 1952
30. Vetterli, M., Kovacevic, J.: Wavelets and Subband Coding. Prentice-Hall, Englewood Cliffs (1995)
31. WebM Project, "VP9 Bitstream & Decoding Process Specification" (v. 0.6), March 2016. http://www.webmproject.org

# Automatic Tongue Recognition Based on Color and Textural Features

Ryszard S. Choraś[✉]

Institute of Telecommunications and Computer Sciences,
UTP University of Science and Technology, S. Kaliskiego 7,
85-796 Bydgoszcz, Poland
choras@utp.edu.pl

**Abstract.** This paper proposes a method of tongue recognition. Tongue images have many advantage for personal identification and verification. In this paper a tongue features are extracted based on color and texture features. These features can be used in forensic applications and with other robust biometrics features can be combined in multi modal biometric system.

## 1 Introduction

Personal identification is crucially significant in a variety of applications. Conventional person's identification systems used:

– Something you have:
  • Token: key, card, or badge
– Something you know:
  • Password
  • PIN numbers
– Something you are:
  • Biometric
    – Physiological
    – Behavioral

A password for personal identification has the risk that the user forgets it or the other persons use it. User's have problems in terms of theft, loss, and reliance on the own memory.

Images play an important role in the identification process of people [2,4]. Biometric identification systems are systems that use pattern recognition ways to identify a specific person by establishing the authenticity of a specific physiological or behavioral characteristic of that person [8].

Biometric systems have four main components: sensor, feature extraction, biometric database, matching-score and decision-making modules. The input subsystem consists of a special sensor needed to acquire the biometric signal. Invariant features are extracted from the signal for representation purposes in the feature extraction subsystem. During the enrollment process, a representation

© Springer International Publishing AG 2017
R.S. Choraś (ed.), *Image Processing and Communications Challenges 8*,
Advances in Intelligent Systems and Computing 525, DOI 10.1007/978-3-319-47274-4_2

(called template) of the biometrics in terms of these features is stored in the system. The matching subsystem accepts query and reference templates and returns the degree of match or mismatch as a score, i.e., a similarity measure. A final decision step compares the score to a decision threshold to the comparison a match or non-match.

A.K. Jain et al. [8] also defines the following requirements that a given measure must satisfy to be a biometric: The ideal biometric characteristics have five qualities:

1. Robust: Unchanging on an individual over time. "Robustness" is measured by the probability that a submitted sample will not match the enrollment image.
2. Distinctive: Showing great variation over the population. "Distinctiveness" is measured by the probability that a submitted sample will match the enrollment image of another user.
3. Available: The entire population should ideally have this measure in multiples. "Availability" is measured by the probability that a user will not be able to supply a readable measure to the system upon enrollment.
4. Accessible: Easy to image using electronic sensors. "Accessibility" can be quantified by the number of individuals that can be processed in a unit time, such as a minute or an hour.
5. Acceptable: People do not object to having this measurement taken on them. "Acceptability" is measured by polling the device users.

Tongue recognition is attracting a great deal of attention because of its usefulness in many applications [3]. Traditional, tongue recognition are often classified into two groups:

– Tongue recognition and analysis for the patient disease diagnosis. Tongue recognition for diagnosis has played an important role in traditional Chinese medicine (*TCM*) and in this area most investigation has been focused on extraction of chromatic features [10,18], shape and textural features [5,6,11].
– Tongue recognition for biometric personal identification.

Our work concerns the biometric applications of the tongue recognition and efficient feature extraction.

Tongue image analysis have received much attention in image analysis and computer vision. Tongue texture has many advantages for human identification and verification [9,20]. The identification of people can be based on the texture features. As a biometric identifier, tongue image has the following properties:

– Tongue images are unique to every person. Texture features of the tongue are distinctive to each person,
– Texture features of an individual tongue are stable and unchangeable during the life of a person,
– The human tongue is well protected in mouth and is difficult to forge.

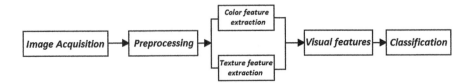

**Fig. 1.** Tongue recognition system

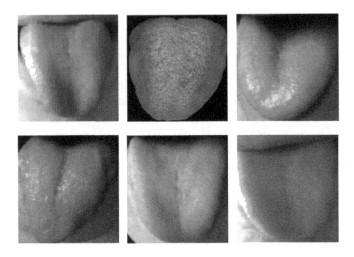

**Fig. 2.** Tongue images

Tongue recognition system is presented in Fig. 1 and it involves five major modules: tongue image acquisition, preprocessing, tongue feature extraction, visual features and classification.

Images which are considered in this paper are displayed in Fig. 2.

## 2   Preprocessing

Before performing feature extraction, the original tongue images are subjected to some image processing operations, such as:

1. Color conversion. The extraction of color features can be performed in different color spaces [17]. Each image is represented using three components of the color space. A color transformation that reduces the psychovisual redundancy and correlation of the image is highly desired. The $YC_rC_b$ is an encoded non-linear $RGB$ signal for image processing work. Color is represented by luminance, computed from nonlinear $RGB$ [14], constructed as a weighted sum of the $RGB$ values, and two color difference values $C_r$ and $C_b$ that are formed by subtracting luminance from $RGB$ red and blue components. The two color

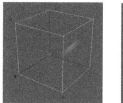

**Fig. 3.** $RGB$ and $YC_bC_r$ color spaces

spaces i.e. $RGB$ and $YC_bC_r$ were used for extraction color features (Fig. 3).

$$Y = 0,299R + 0,587G + 0,114B$$
$$C_r = 0,713(R - Y)$$
$$C_b = 0,564(B - Y)$$
(1)

or

$$\begin{bmatrix} Y \\ C_r \\ C_b \end{bmatrix} = \begin{bmatrix} 0,299 & 0,587 & 0,114 \\ 0,500 & -0,419 & -0,081 \\ -0,169 & -0,331 & 0,500 \end{bmatrix} \cdot \begin{bmatrix} R \\ G \\ B \end{bmatrix}$$
(2)

An image histogram refers to the probability mass function of the image intensities. This is extended for color images to capture the joint probabilities of the intensities of the three color channels. More formally, the color histogram is defined by,

$$h_{A,B,C} = N \cdot Prob(A = a, B = b, C = c)$$
(3)

where $A$, $B$ and $C$ represent the three color channels ($R,G,B$ or $YC_rC_b$) and $N$ is the number of pixels in the image. Computationally, the color histogram is formed by discretizing the colors within an image and counting the number of pixels of each color.

2. Extraction of region of interest *(ROI)* from original tongue images. The tongue images are normalized with respect to position, orientation, scale, reflection, as follows.

The new invariant coordinates $(x, y)$ of image pixels and the old coordinates $(x', y')$ are related by

$$[x, y, 1] = [x', y'1] \times \begin{bmatrix} 1 & 0 & 0 \\ 0 & 1 & 0 \\ -i_0 & -j_0 & 1 \end{bmatrix}$$
$$\times \begin{bmatrix} \frac{1}{\delta_x} & 0 & 0 \\ 0 & \frac{1}{\delta_y} & 0 \\ 0 & 0 & 1 \end{bmatrix} \times \begin{bmatrix} \cos\beta & \sin\beta & 0 \\ -\sin\beta & \cos\beta & 0 \\ 0 & 0 & 1 \end{bmatrix}$$
(4)

where $x_0, y_0$ is the centroid of image; $\delta_x$ and $\delta_y$ represent standard deviation relative to variable $x, y$; and $\beta$ is an angle between the major axis of an object and the vertical line

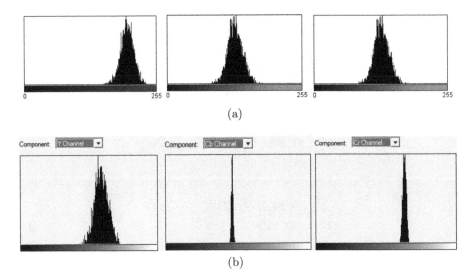

(a)

(b)

**Fig. 4.** Histograms: (a) $RGB$ channels and (b) $YC_bC_r$ channels

$$\tan 2\beta = \frac{2\sum_x \sum_y (x - x_0)(y - y_0)}{\sum_x (x - x_0^2) - \sum_y (y - y_0^2)} \tag{5}$$

Next, the $ROI$'s tongue blocks are automatically selected on the centroid of tongue normalized images. The size of whole $ROI$ is $w_x \times w_y$ where $w_x = (x_0 + \frac{K}{2}) - (x_0 - \frac{K}{2})$, $w_y = (y_0 + \frac{K}{2}) - (y_0 - \frac{K}{2})$ where $K = 128$ pixels (Fig. 5). Next, the $ROI$ image is divided into the four sub-blocks. The size of sub-block is $\frac{K}{2} \times \frac{K}{2}$ pixels (Fig. 6).

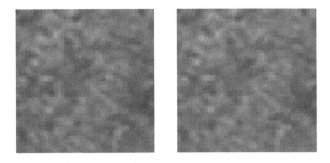

**Fig. 5.** Tongue $ROI$

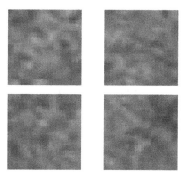

**Fig. 6.** Four $ROI$'s sub-blocks

# 3 Feature Extraction

## 3.1 Feature Extraction Based on Color Moments

The distribution of color was represented by color histograms (Fig. 4), and formed the image's feature vectors. The mathematical foundation of this approach is that any probability distribution is uniquely characterized by its moments.

Color moments have been successfully used in many image processing/biometrics systems. The first order (mean), the second (variance), the third order (skewness) and the fourth order (kurtosis) color moments have been proved to be efficient and effective in representing color distributions of images [16]. Mathematically, the first three moments are defined

$$\mu_c = \frac{1}{MN} \sum_{x=1}^{M} \sum_{y=1}^{N} f_c(x,y) \tag{6}$$

$$\sigma_c = \left(\frac{1}{MN} \sum_{x=1}^{M} \sum_{y=1}^{N} (f_c(x,y) - \mu_c)^2\right)^{\frac{1}{2}} \tag{7}$$

$$s_c = \frac{1}{MN} \sum_{x=1}^{M} \sum_{y=1}^{N} \left[\frac{f_c(x,y) - \mu_c}{\sigma_c}\right]^3 \tag{8}$$

$$k_c = \left\{\frac{1}{MN} \sum_{x=1}^{M} \sum_{y=1}^{N} \left[\frac{f_c(x,y) - \mu_c}{\sigma_c}\right]^4\right\} - 3 \tag{9}$$

where $f_c(x,y)$ is the value of the $c$-th color component of the image pixel $(x,y)$, and $MN$ is the number of pixels in the image.

The color features are computed in $RGB$ and $YC_bC_r$ color spaces.

Since only 24 (four moments for each of the three color components in two color spaces) numbers are used to represent the color content of each image, color moments are a very compact representation compared to other color features (Table 1).

**Table 1.** Moments of color components

| Image | Color components | Mean $\mu_c$ | Variance $\sigma_c$ | Skewness $s_c$ | Kurtosis $k_c$ |
|-------|------------------|--------------|---------------------|----------------|----------------|
| Tongue ROI Fig. 5 | R | 198,72 | 13,07 | 0,201 | 0.267 |
| | G | 129,14 | 14,39 | 0,889 | 0,158 |
| | B | 132,95 | 14,44 | −0,020 | −0,013 |
| Tongue ROI Fig. 5 | Y | 149,81 | 13,63 | −0,030 | 0,184 |
| | $C_b$ | 117,71 | 1,92 | −0,074 | 0,389 |
| | $C_r$ | 161,99 | 3,454 | −0,074 | −0,3 |

## 3.2   Gabor Filters for Feature Extraction

Gabor filters are a powerful tool to extract texture features and in the spatial domain is a complex exponential modulated by a Gaussian function. In the most general the Gabor filters are defined as follows [1,13,15].

The two-dimensional Gabor filter is defined as

$$Gab(x,y,W,\theta,\sigma_x,\sigma_y) = \frac{1}{2\pi\sigma_x\sigma_y}e^{\left[-\frac{1}{2}\left(\left(\frac{x}{\sigma_x}\right)^2+\left(\frac{y}{\sigma_y}\right)^2\right)+jW(x\cos\theta+y\sin\theta)\right]} \quad (10)$$

where $j = \sqrt{-1}$ and $\sigma_x$ and $\sigma_y$ are the scaling parameters of the filter, $W$ is the radial frequency of the sinusoid and $\theta \in [0, \pi]$ specifies the orientation of the Gabor filters [7].

Figure 7 presents the real and imaginary parts of Gabor filters.

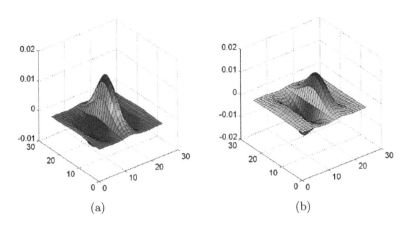

(a)                              (b)

**Fig. 7.** The real and imaginary parts of Gabor filters

In our work we use a bank of filters built from the real part of Gabor expression called as even symmetric Gabor filter. Gabor filtered output of the image is obtained by the convolution of the image with Gabor even function for each of the orientation/spatial frequency (scale) orientation (Fig. 8).

Given an image $F(x, y)$, we filter this image with $Gab(x, y, W, \theta, \sigma_x, \sigma_y)$

$$FGab(x, y, W, \theta, \sigma_x, \sigma_y) = \sum_k \sum_l F(x - k, y - l) * Gab(x, y, W, \theta, \sigma_x, \sigma_y) \quad (11)$$

The magnitudes of the Gabor filters responses are represented by three moments

$$\mu(W, \theta, \sigma_x, \sigma_y) = \frac{1}{XY} \sum_{x=1}^{X} \sum_{y=1}^{Y} FGab(x, y, W, \theta, \sigma_x, \sigma_y) \quad (12)$$

$$std(W, \theta, \sigma_x, \sigma_y) = \sqrt{\sum_{x=1}^{X} \sum_{y=1}^{Y} ||FGab(x, y, W, \theta, \sigma_x, \sigma_y)| - \mu(W, \theta, \sigma_x, \sigma_y)|^2} \quad (13)$$

$$Energy = \sum_{x=1}^{X} \sum_{y=1}^{Y} [FGab(x, y, W, \theta, \sigma_x, \sigma_y)]^2 \quad (14)$$

By selecting different center frequencies and orientations, we can obtain a family of Gabor kernels, which can then be used to extract features from an image. The feature vector is constructed using *mean* - $\mu(W, \theta, \sigma_x, \sigma_y)$, *standard deviation* - $std(W, \theta, \sigma_x, \sigma_y)$ and *energy* as feature components (Table 2).

We defined the vectors of features as follows:

$$FV = (Feature_{Color\ moments}, Feature_{Gabor}) \quad (15)$$

The first part of the $FV$ contains the 24 color moments. The features in second part of $FV$ are listed as follows $Feature_{Gabor} = ((\mu_1(x, y), std_1(x, y), Skew_1) \ldots (\mu_t(x, y), std_t(x, y), Skew_t))$.

To reduce dimension of feature vector [12, 19], we use the Principle Component Analysis ($PCA$) algorithm to keep the most useful Gabor features.

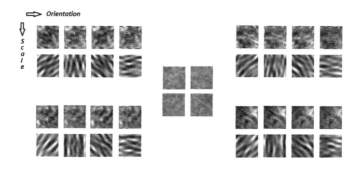

**Fig. 8.** Gabor images of tongue $ROI$'s

**Table 2.** Features of tongue *ROI*'s.

ROI 1

| Scale | Orientation | Energy | Mean | Std |
|-------|-------------|--------|------|-----|
| 2 | 45° | 1968755.0 | 480.65308 | 35.064735 |
| 2 | 90° | 1999251.1 | 488.09842 | 40.68314 |
| 2 | 135° | 1968758.9 | 480.65402 | 34.23929 |
| 2 | 180° | 1999252.2 | 488.0987 | 42.19372 |
| 8 | 45° | 3.181323E7 | 7766.903 | 669.66876 |
| 8 | 90° | 3.1827126E7 | 7770.2944 | 657.5264 |
| 8 | 135° | 3.181321E7 | 7766.897 | 762.5107 |
| 8 | 180° | 3.1827142E7 | 7770.2983 | 728.3237 |

ROI 2

| Scale | Orientation | Energy | Mean | Std |
|-------|-------------|--------|------|-----|
| 2 | 45° | 1968755.0 | 480.65308 | 35.064735 |
| 2 | 90° | 1999251.1 | 488.09842 | 40.68314 |
| 2 | 135° | 1968758.9 | 480.65402 | 34.23929 |
| 2 | 180° | 1999252.2 | 488.0987 | 42.19372 |
| 8 | 45° | 3.181323E7 | 7766.903 | 669.66876 |
| 8 | 90° | 3.1827126E7 | 7770.2944 | 657.5264 |
| 8 | 135° | 3.181321E7 | 7766.897 | 762.5107 |
| 8 | 180° | 3.1827142E7 | 7770.2983 | 728.3237 |

ROI 3

| Scale | Orientation | Energy | Mean | Std |
|-------|-------------|--------|------|-----|
| 2 | 45° | 1867638.8 | 455.9665 | 36.981506 |
| 2 | 90° | 1896568.8 | 463.02948 | 44.25503 |
| 2 | 135° | 1867638.6 | 455.96646 | 36.304142 |
| 2 | 180° | 1896567.2 | 463.0291 | 41.741203 |
| 8 | 45° | 3.0179278E7 | 7367.988 | 707.88116 |
| 8 | 90° | 3.0192456E7 | 7371.205 | 717.7458 |
| 8 | 135° | 3.0179264E7 | 7367.984 | 774.104 |
| 8 | 180° | 3.0192524E7 | 7371.221 | 614.612 |

ROI 4

| Scale | Orientation | Energy | Mean | Std |
|-------|-------------|--------|------|-----|
| 2 | 45° | 1906920.6 | 465.5568 | 27.945246 |
| 2 | 90° | 1936458.5 | 472.7682 | 35.911766 |
| 2 | 135° | 1906919.8 | 465.55658 | 28.149256 |
| 2 | 180° | 1936459.8 | 472.7685 | 35.161488 |
| 8 | 45° | 3.0813984E7 | 7522.9453 | 786.99506 |
| 8 | 90° | 3.08275E7 | 7526.245 | 728.3674 |
| 8 | 135° | 3.0814002E7 | 7522.9497 | 562.7714 |
| 8 | 180° | 3.082751E7 | 7526.2476 | 521.4142 |

Let $X = [x_1, x_2, \ldots, x_n]$ denote an $n$-dimensional feature vector. The mean of the vector $X$ and the total scatter covariance matrix of the vector $X$ are defined as: $\overline{\mu} = \frac{1}{n} \sum_{i=1}^{n} x_i$ and $S_X = \sum_{i=1}^{n} (x_i - \overline{\mu}) \cdot (x_i - \overline{\mu})^t$.

The $PCA$ projection matrix $S$ can be obtained by eigen-analysis of the covariance matrix $S_X$. We compute the eigenvalues of $S_X : \lambda_1 > \lambda_2 > \cdots > \lambda_n$ and the eigenvectors of $S_X : s_1, s_2, \ldots, s_n$. Thus $S_X s_i = \lambda_i s_i, \ i = 1, 2, \ldots, m$. $s_i$ is the $i$th largest eigenvector of $S_X$, $m \ll n$ and $S = [s_1, s_2, \ldots, s_m]$.

Any vector $x$ can be written as a linear combination of the eigenvectors ($S$ is symmetric, $s_1, s_2, \ldots, s_n$ form a basis), i.e. $x = \sum_{i=1}^{n} b_i u_i$. For dimensionality reduction we choose only $m$ largest eigen values, i.e. $x = \sum_{i=1}^{m} b_i u_i$. $m$ is choose as follows: $\frac{\sum_{i=1}^{m} \lambda_i}{\sum_{i=1}^{n} \lambda_i} > t$ where $t$ is threshold.

By removing the principal components that contribute little to the variance, we project the entire feature vector to a lower dimensional space, but retain most of the information.

## 4   Conclusion

In the paper, are presented some approaches for tongue recognition from images. To evaluate the performance of tongue recognition methods we use own tongue database that consists 30 images. We proposed a method which combines the recognition results of Gabor filters and color moments features to tongue recognition. The proposed system will be evaluated on other tongue databases in the future study.

**Acknowledgment.** The research was supported by the UTP University of Sciences and Technology by the Grant BS01/2014.

## References

1. Choras, R.S.: Iris-based person identification using Gabor wavelets and moments. In: Proceedings of 2009 International Conference on Biometrics and Kansei Engineering ICBAKE, pp. 55–59. CPS IEEE Computer Society (2009)
2. Choras, R.S.: Thermal Face Recognition, Image Processing & Communications Challenges 7. AISC, vol. 389, pp. 37–46. Springer, Berlin (2016)
3. Choras, R.S.: Tongue recognition from images. In: Proceedings of the First International Conference on Advances in Signal, Image and Video Processing SIGNAL 2016, Lisbon, Portugal, pp. 6–11 (2016)
4. Choras, R.S.: Vascular Biometry, Image Processing & Communications Challenges 6. AISC, vol. 313, pp. 21–28. Springer, Berlin (2015)
5. Huang, W., Yan, Z., Xu, J., Zhang, L.: Analysis of the tongue fur and tongue features by naive bayesian classifier. In: 2010 International Conference on Computer Application and System Modeling (ICCASM 2010), vol. 4, pp. V4–304, IEEE, October 2010
6. Huang, B., Zhang, D., Li, Y., Zhang, H., Li, N.: Tongue coating image retrieval. In: Proceedings of the 3rd International Conference on Advanced Computer Control, Harbin, pp. 292–296 (2011)

7. Jain, A.K.: Fundamentals of a Digital Signal Processing. Prentice-Hall, Englewood Clifts (1989)
8. Jain, A.K., Bolle, R., Pankanti, S.: Biometrics: Personal Identification in Networked Society. Kulwer Academic, London (1999)
9. Lahmiri, S.: Recognition of tongueprint textures for personal authentication: a wavelet approach. J. Adv. Inf. Technol. **3**(3), 168–175 (2012)
10. Li, C., Yuen, P.: Tongue image matching using color content. Pattern Recogn. **35**(2), 407–419 (2002)
11. Li, W., Hu, S., Yao, J., Song, H.: The separation framework of tongue coating and proper in traditional Chinese medicine. In: Proceedings of the 7th International Conference Information, Communications and Signal Processing, Macau, December, pp. 1–4 (2009)
12. Liu, C.J., Wechsler, H.: Gabor feature based classification using the enhanced fisher linear discriminant model for face recognition. IEEE Trans. Image Process. **11**(4), 467–476 (2002)
13. Liu, D.H., Lam, K.M., Shen, L.S.: Optimal sampling of gabor features for face recognition. Pattern Recogn. Lett. **25**(2), 267–276 (2004)
14. Poynton, C.A.: Frequently asked questions about colour (1995). ftp://www.inforamp.net/pub/users/poynton/doc/colour/
15. Shen, L., Bai, L.: Face recognition based on gabor features using kernel methods. In: Proceedings 6th IEEE Conference on Face and Gesture Recognition, Korea, pp. 170–175 (2004)
16. Stricker, M., Orengo, M.: Similarity of color images. In: SPIE Storage and Retrieval for Image and Video Databases III, vol. 2185, pp. 381–392 (1995)
17. Swain, M.J., Ballard, D.H.: Color indexing. Int. J. Comput. Vis. **7**(1), 11–32 (1991)
18. Wang, Y., Yang, J., Zhou, Y.: Region partition and feature matching based color recognition of tongue image. Pattern Recogn. Lett. **28**(1), 11–19 (2007)
19. Zhang, B.C., Shan, S.G., Chen, X.L., Gao, W.: Histogram of Gabor phase patterns (HGPP): a novel object representation approach for face recogniton. IEEE Trans. Image Process. **16**, 57–68 (2007)
20. Zhang, D., Liu, Z., Yan, J., Shi, P.: Tongue-print: a novel biometrics pattern. In: Lee, S.-W., Li, S.Z. (eds.) ICB 2007. LNCS, vol. 4642, pp. 1174–1183. Springer, Heidelberg (2007). doi:10.1007/978-3-540-74549-5_122

# A First Attempt to Construct Effective Concept Drift Detector Ensembles

Michał Woźniak$^{(\boxtimes)}$, Paweł Ksieniewicz, Andrzej Kasprzak, Karol Puchała, and Przemysław Ryba

Faculty of Electronics, Department of Systems and Computer Networks, Wroclaw University of Science and Technology, Wroclaw, Poland
michal.wozniak@pwr.edu.pl

**Abstract.** The big data is usually described by so-called 5Vs (Volume, Velocity, Variety, Veracity, Value). The business success in the big data era strongly depends on the smart analytical software which can help to make efficient decisions (Value for enterprise). Therefore, the decision support software should take into consideration especially that we deal with massive data (Volume) and that data usually comes continuously in the form of so-called *data stream* (Velocity). Unfortunately, most of the traditional data analysis methods are not ready to efficiently analyze fast growing amount of the stored records. Additionally, one should also consider phenomenon appearing in data stream called *concept drift*, which means that the parameters of an using model are changing, what could dramatically decrease the analytical model quality. This work is focusing on the classification task, which is very popular in many practical cases as fraud detection, network security, or medical diagnosis. We propose how to detect the changes in the data stream using combined concept drift detection model. The experimental evaluations show that it is an interesting direction, what encourage us to use it in practical applications.

**Keywords:** Data stream · Concept drift · Pattern classification · Drift detector

## 1 Introduction

The analysis of streaming data is recently the focus of intense research, because the contemporary companies desire effective analytical tools which are able to deliver valuable information to build the competitive advantage of a given company. However, most of the traditional classifier design methods do not take into consideration that dependencies between feature values described incoming objects and their labels may change and that labeling each of the arriving example is impossible from the practical point of view.

Let's focus on the first problem called *concept drift* [11] and it comes in many forms, depending on the type of change. Appearance of concept drift may spoil quality of used models. Therefore, developing positive methods which are

© Springer International Publishing AG 2017
R.S. Choraś (ed.), *Image Processing and Communications Challenges 8*,
Advances in Intelligent Systems and Computing 525, DOI 10.1007/978-3-319-47274-4_3

able to effectively deal with this phenomena has become an increasing issue. We may distinguish two types of drifts according their influences into probabilistic characteristics of a classification task [4]:

- *virtual concept drift* means that the changes do not have any impact on decision boundaries (some works report that they do not have an impact on *posterior* probability, but it is disputable) [10].
- *real concept drift* means that the changes have an impact on decision boundaries [11].

We may also focus on drift's impetuosity:

- *gradual or incremental drift* - for the gradual drift for a given period of time examples from different model could appear in the steam concurrently, while for incremental drift the model's parameters are changing smoothly.
- *sudden drift*, where the drift has rapid nature.

Concept drift detector is an algorithm, which on the basis of information about new incoming examples and model's performance is able to return information that data stream distributions are changing. The detectors inform about drift detection or warning level. The detection signal causes the used model should be rebuilt as quick as possible, while achieving warning level enforces that new data set should be collected for future model updating.

The drift detection could be recognized as the simple classification task, but from practical point of view detectors do not use the classical classification model, because they rather solve the regression problem than classification one. The detection is hard, because on the one hand we require concept drift detection as soon as possible to replace outdated model and to reduce so-called *restoration time*, but on the other hand we do not accept to many false alarms [6]. Therefore to measure performance of concept drift detectors the following metrics are usually used:

- Number of correct detected drifts.
- Number of false alarms.
- Time between real drift appearance and its detection.

In some works, aggregated measures, which take into consideration the mentioned above metrics, are proposed [2], but we decided not to use them because using aggregated measures do not allow to precisely analyse behavior of the considered detectors.

It is worth noticing that the detectors are assuming the continue access to class labels, which usually cannot be granted from the practical point of view. Therefore, during constructing the concept drift detectors we have to take into consideration the cost of data labeling, which is usually passed over. It seems to be very interesting to design detectors employing (called *active learning*) paradigm [5] or unlabeled examples only. Unfortunately, it is easy to show that without access to class labels the real drift could be undetected [9].

## 2    Combined Concept Drift Detectors

Let's assume that we have a pool of $n$ drift detectors

$$\Pi = \{D_1, D_2, ..., D_n\} \tag{1}$$

Each of them returns signal that warning level is achiever or concept drift is detected, i.e.,

$$D_i = \begin{cases} 0 \text{ if drift is not detected} \\ 1 \text{ if warning level is achieved} \\ 2 \text{ if drift is detected} \end{cases} \tag{2}$$

As yet not so many papers deal with combined drift detectors. Bifet et al. [2] proposed the simple combination rules based on the appearance of drift once ignoring signals about warning level.

Let's present the combination rules which allow to make a decision on the basis of individual concept drift detector outputs. Because this work focuses on the ensemble of heterogeneous detectors, therefore only the combination rules using drift appearance signal are taken into consideration.

### 2.1    ALO (*At Least One detects drift*)

Combined detector makes a decision about drift if at least one individual detector returns decision that drift appears.

$$ALO(\Pi, x) = \begin{cases} 0 & \sum_{i=1}^{n}[D_i(x) = 2] \\ 1 & \sum_{i=1}^{n}[D_i(x) = 0] \end{cases} \tag{3}$$

where [] denotes Iverson bracket.

### 2.2    ALHWD (*At Least Half of the Detectors returns Warnings or detect Drift*)

Detector ensemble makes a decision about drift if at half of individual detectors return decisions that drift appears or warning level is achieved.

$$ALHWD(\Pi, x) = \begin{cases} 0 & \sum_{i=1}^{n}[D_i(x) > 0] < \frac{n}{2} \\ 1 & \sum_{i=1}^{n}[D_i(x) > 0] \geqslant \frac{n}{2} \end{cases} \tag{4}$$

### 2.3    ALHD (*At Least Half of the detectors detect Drift*)

Committee of detectors makes a decision about drift if at half of individual detectors return decisions that drift appears.

$$ALHWD(\Pi, x) = \begin{cases} 0 & \sum_{i=1}^{n}[D_i(x) = 2] < \frac{n}{2} \\ 1 & \sum_{i=1}^{n}[D_i(x) = 2] \geqslant \frac{n}{2} \end{cases} \tag{5}$$

## 2.4   AWD (*All detectors return Warnings or detect Drift*)

Combined detector makes a decision about drift if each individual detector returns decisions that drift appears or warning level is achieved.

$$ALHWD(\Pi, x) = \begin{cases} 0 & \sum_{i=1}^{n}[D_i(x) > 0] < n \\ 1 & \sum_{i=1}^{n}[D_i(x) > 0] \geqslant n \end{cases} \tag{6}$$

## 2.5   AD (*All detectors detect Drift*)

Committee of detectors makes a decision about drift if each individual detector returns decisions that drift appears.

$$ALHWD(\Pi, x) = \begin{cases} 0 & \sum_{i=1}^{n}[D_i(x) = 2] < n \\ 1 & \sum_{i=1}^{n}[D_i(x) = 2] \geqslant n \end{cases} \tag{7}$$

Let's notice that proposed combined detectors are *de facto* classifier ensembles [12], which use deterministic combination rules, which do not take into consideration any additional information, e.g., about individual drift detector's qualities. In this work, we also do not focus on the very important problem how choose the valuable pool of individual detectors. For classifier ensemble such a process (called ensemble selection or ensemble pruning) uses diversity measure [8], but for the combined detectors the measures have been using thus far is impossible, because of different nature of the decision task.

# 3   Experimental Research

## 3.1   Goals

The main objective of the experimental study was evaluating the proposed combined concept drift detectors and their comparison with the well-known simple methods. To ensure the appropriate diversity of the pool of detectors we decided to produce an ensemble on the basis of five detectors employing different models presented in the previous section. For each experiment we estimated detector sensitivity, number of false alarms and computational complexity of the model, i.e., commutative running time.

## 3.2   Set-Up

We used the individual detectors based on McDiarmid's inequality analysis [3] with the following parameters:

– HDDM_A 0.001 (drift confidence 0.001, warning confidence 0.005)
– HDDM_A 0.002 (drift confidence 0.002, warning confidence 0.006)
– HDDM_A 0.003 (drift confidence 0.003, warning confidence 0.007)
– HDDM_A 0.004 (drift confidence 0.004, warning confidence 0.008)
– HDDM_A 0.005 (drift confidence 0.005, warning confidence 0.009).

For each detectors lambda was equal 0.05 and one-side t-test was used. To build the detector ensemble the five proposed in the previous section combination rules were used. All experiments were carried out using MOA (*Massive Online Analysis*)[1] and our own software written in Java according to MOA's requirements [1].

For each experiment 3 computer generated data streams were used. Each of them consists of 10 000 examples:

– Data stream with sudden drift appearing after each 5.000 examples.
– Data stream with gradual drift, where 2 concept appear.
– Data stream without drift.

The stationary data stream (without drift) was chosen because we would like to evaluate the sensitivity of the detector, i.e., number of false alarms.

### 3.3   Results

The results of experiment were presented in Tables 1, 2 and 3[2].

**Table 1.** Results of experiment for data stream with sudden drift

| Detector | No. of real drifts | No. of corrected detected drifts | Average detection delay | No. of false detections | Time |
|---|---|---|---|---|---|
| HDDM_A 0.001 | 20 | 20 | 11,65 | 0 | 1,69 |
| HDDM_A 0.002 | 20 | 20 | 10,35 | 0 | 1,8 |
| HDDM_A 0.003 | 20 | 20 | 9,8 | 0 | 1,62 |
| HDDM_A 0.004 | 20 | 20 | 8,95 | 0 | 1,68 |
| HDDM_A 0.005 | 20 | 20 | 8,6 | 0 | 1,66 |
| ALO min | 20 | 20 | 4,15 | 517 | 8,48 |
| ALHWD | 20 | 20 | 80,9 | 29 | 8,68 |
| ALHD mv | 20 | 19 | 127,5 | 19 | 5,51 |
| AWD | 20 | 0 | – | – | 8,92 |
| AD max | 20 | 0 | – | – | 5,8 |

### 3.4   Discussion

Firstly we have to emphasize that we realize that the scope of the experiments was limited therefore drawing the general conclusions is very risky. The results are not so promising, especially for sudden drift, where sensitive rules as ALO, ALHWD and ALHD detected drifts correctly, but returned so many false alarms.

---

[1]  http://moa.cms.waikato.ac.nz/.
[2]  The detailed results of the experiments could be found https://drive.google.com/drive/u/0/folders/0B8ja_TIQel7KSzdLYkVGLVoySG8.

**Table 2.** Results of experiment for data stream with gradual drift

| Detector | No. of real drifts | No. of corrected detected drifts | Average detection delay | No. of false detections | Time |
|---|---|---|---|---|---|
| HDDM_A 0.001 | 1 | 1 | 242 | 1 | 1,77 |
| HDDM_A 0.002 | 1 | 1 | 241 | 1 | 1,68 |
| HDDM_A 0.003 | 1 | 1 | 141 | 2 | 1,7 |
| HDDM_A 0.004 | 1 | 1 | 134 | 2 | 1,67 |
| HDDM_A 0.005 | 1 | 1 | 134 | 2 | 1,63 |
| ALO min | 1 | 1 | 2 | 75 | 8,9 |
| ALHWD | 1 | 1 | 107 | 4 | 8,9 |
| ALHD mv | 1 | 1 | 180 | 3 | 5,8 |
| AWD | 1 | 0 | – | – | 8,78 |
| AD max | 1 | 0 | – | – | 5,73 |

**Table 3.** Results of experiment for data stream without drift

| Detector | No. of real drifts | No. of corrected detected drifts | Average detection delay | No. of false detections | Time |
|---|---|---|---|---|---|
| HDDM_A 0.001 | 0 | 0 | – | 0 | 1,7 |
| HDDM_A 0.002 | 0 | 0 | – | 0 | 1,64 |
| HDDM_A 0.003 | 0 | 0 | – | 1 | 1,64 |
| HDDM_A 0.004 | 0 | 0 | – | 1 | 1,61 |
| HDDM_A 0.005 | 0 | 0 | – | 1 | 1,66 |
| ALO min | 0 | 0 | – | 498 | 8,58 |
| ALHWD | 0 | 0 | – | 35 | 8,95 |
| ALHD mv | 0 | 0 | – | 39 | 5,47 |
| AWD | 0 | 0 | – | 0 | 8,5 |
| AD max | 0 | 0 | – | 0 | 5,34 |

For gradual drift only ALHWD and ALHD behaved quite well, but number of false alarms was still higher than simple methods. On the basis of the experiment we are not able to say expressly if the combined concept drift detectors are promising direction. Nevertheless, we decided to continue the works on such models, especially for a pool of heterogeneous detectors and method which are able to prune the detector ensemble. We have also to notice the main drawback of the proposed models, because they are more complex than simple ones, but using the proposed method of parallel interconnection they are easy to parallelize and could be run in a distributed computing environment.

# 4   Final Remarks

In this work five deterministic combination rules for combined concept drift detectors were discussed. They seem to be an interesting proposition to solve the problem of concept drift detection, nevertheless the results of experiments do not confirm their high quality, but as we mentioned above it is probably caused by very naive choice of the individual models to the detector ensemble.

Let's propose the future research directions:

- Developing methods how to choose individual detectors to an ensemble.
- Proposing cost effective method of drift detection, i.e., following the observation that usually we are not able to ensure the access to class label for each incoming example an semi-supervised and unsupervised methods of combined detector training seem to be very interesting.
- Proposing new trained combination rules to fully exploit the strengths of the individual detectors.
- Developing the combined local drift detectors, e.g. based on the AdaSS algorithm [7], because many changes have the local nature.

**Acknowledgements.** This work was supported by the statutory funds of the Department of Systems and Computer Networks, Faculty of Electronics, Wroclaw University of Science and Technology and by the Polish National Science Centre under the grant no. DEC-2013/09/B/ST6/02264. All computer experiments were carried out using computer equipment sponsored by EC under FP7, Coordination and Support Action, Grant Agreement Number 316097, ENGINE European Research Centre of Network Intelligence for Innovation Enhancement (http://engine.pwr.edu.pl/).

# References

1. Bifet, A., Holmes, G., Kirkby, R., Pfahringer, B.: Moa: massive online analysis. J. Mach. Learn. Res. **11**, 1601–1604 (2010). http://dl.acm.org/citation.cfm?id=1756006.1859903
2. Bifet, A., Read, J., Pfahringer, B., Holmes, G., Žliobaitė, I.: CD-MOA: change detection framework for massive online analysis. In: Tucker, A., Höppner, F., Siebes, A., Swift, S. (eds.) IDA 2013. LNCS, vol. 8207, pp. 92–103. Springer, Heidelberg (2013). doi:10.1007/978-3-642-41398-8_9
3. Blanco, I.I.F., del Campo-Avila, J., Ramos-Jimenez, G., Bueno, R.M., Diaz, A.A.O., Mota, Y.C.: Online and non-parametric drift detection methods based on hoeffding's bounds. IEEE Trans. Knowl. Data Eng. **27**(3), 810–823 (2015). http://dx.doi.org/10.1109/TKDE.2014.2345382
4. Gama, J., Zliobaite, I., Bifet, A., Pechenizkiy, M., Bouchachia, A.: A survey on concept drift adaptation. ACM Comput. Surv. (CSUR), **46**(4) (2014). Surveys Homepage archive. Article No. 44
5. Greiner, R., Grove, A.J., Roth, D.: Learning cost-sensitive active classifiers. Artif. Intell. **139**(2), 137–174 (2002)
6. Gustafsson, F.: Adaptive Filtering and Change Detection. Wiley, October 2000. http://www.wiley.com/WileyCDA/WileyTitle/productCd-0471492 876,descCd-description.html

7. Jackowski, K., Krawczyk, B., Wozniak, M.: Adass+ the hybrid training method of a classifier based on a feature space partitioning. Int. J. Neural Syst. **24**(3), 1430007 (2014)
8. Kuncheva, L.I.: Combining Pattern Classifiers: Methods and Algorithms. Wiley-Interscience, Chichester (2004)
9. Sobolewski, P., Wozniak, M.: Concept drift detection and model selection with simulated recurrence and ensembles of statistical detectors. J. Univ. Comput. Sci. **19**(4), 462–483 (2013)
10. Widmer, G., Kubat, M.: Effective learning in dynamic environments by explicit context tracking. In: Brazdil, P.B. (ed.) ECML 1993. LNCS, vol. 667, pp. 227–243. Springer, Heidelberg (1993). doi:10.1007/3-540-56602-3_139
11. Widmer, G., Kubat, M.: Learning in the presence of concept drift and hidden contexts. Mach. Learn. **23**(1), 69–101 (1996)
12. Wozniak, M., Grana, M., Corchado, E.: A survey of multiple classifier systems as hybrid systems. Inf. Fusion **16**, 3–17 (2014). Special Issue on Information Fusion in Hybrid Intelligent Fusion Systems. http://www.sciencedirect.com/science/article/pii/S156625351300047X

# Quality Prediction of Compressed Images via Classification

Jevgenij Tichonov[1(✉)], Olga Kurasova[1], and Ernestas Filatovas[2]

[1] Institute of Mathematics and Informatics, Vilnius University,
Universiteto str. 3, Vilnius, Lithuania
jevgenij.tichonov@gmail.com, olga.kurasova@mii.vu.lt
[2] Faculty of Fundamental Science, Vilnius Gediminas Technical University,
Sauletekio av. 11, Vilnius, Lithuania
ernest.filatov@gmail.com

**Abstract.** In this paper, we have investigated an image classification problem according to image quality after compression. A classification-based image compression approach has been proposed, where images are assigned to one of two classes before their compression by a JPEG algorithm. This classification allows to set the proper value of Quality Factor (QF) for each image, in such a way, to save storage space while maintaining sufficiently high image quality. The image quality has been evaluated by a Structural Similarity (SSIM) index metric and Peak Signal-to-Noise Ratio (PSNR). As image classification results depend on the selected features describing the images, the feature selection problem has to be solved before classification. The experimental investigation has shown that the proposed approach allows to save storage space compared to a conventional JPEG algorithm. It is especially useful when saving huge amount of images.

**Keywords:** Image quality · Image classification · JPEG algorithm · Quality prediction

## 1 Introduction

Nowadays digital cameras allow to get high quality images which take a lot of storage spaces. Huge sets of high-resolution digital images are generated in medicine, astronautics, social networks, etc. Commonly such images are compressed by lossy compression algorithms that sufficiently reduce image size as well as the storage space, however the image quality also decreases. The JPEG algorithm is widely used for image compression and storage. Digital image processing tools enable to vary many parameters of the JPEG algorithm when selecting the most suitable values for a particular image. Quality Factor (QF) is one of the most important parameter which controls the degree of compression, thereby a quality of the compressed image. It is an integer between 0 and 100 and is used to scale the values in a quantization matrix. However, even in the case of the same value of QF, depending on the image content, the images of different quality and

© Springer International Publishing AG 2017
R.S. Choraś (ed.), *Image Processing and Communications Challenges 8*,
Advances in Intelligent Systems and Computing 525, DOI 10.1007/978-3-319-47274-4_4

sizes are obtained after compression. It is difficult to determine in advance how the values of QF will influence a certain image. Commonly the value selection is performed manually by iteratively compressing an image with different values using image processing tools. Afterwards, the most appropriate resulting image is chosen according to obtained quality and size. This procedure is time consuming, especially when a huge set of images must be processed. Thus, it would be useful to develop an automatic approach that assigns uncompressed images to different classes according to their quality after compression.

In this research, we suggest to predict image quality before usage of a compression algorithm via classification. The classification-based approach for image compression is proposed and investigated, where the goal is to classify images into two classes (high and low quality images after compression) in order to predict the proper value of Quality Factor (QF) for each image to be compressed by the JPEG algorithm.

The remainder of this paper is organized as follows. Section 2 reviews researches on improvement of the JPEG algorithm for more effective image compression. The proposed classification-based image compression approach is described in Sect. 3. The results of the experimental investigation are presented in Sect. 4. Finally, conclusions are drawn in Sect. 5.

## 2    Related Works

The image quality assessment is an integral part of compression applications. Recently a series of quality assessment metrics have been proposed in literature [11]. One of the most widely-used is Peak Signal-to-Noise Ratio (PSNR) [4,12]. The basis of this metrics is Mean Square Error (MSE) [16] between the original (reference) and distorted (compressed) images. However, high PSNR does not guarantee a high image quality [4,16]. Another popular metrics is Structural Similarity (SSIM) index [17] which assesses the visual impact of three image characteristics: luminance, contrast and structure. This metrics is more complex, but accurate, therefore it has become of great interest during the last decade.

As the JPEG algorithm has become the image compression standard [5,14], many efforts have been done by scientists to improve its efficiency. Despite of coming JPEG2000 [13], the JPEG algorithm remains popular, widely-used and deeply-investigated. Classification methods have been applied successfully for improvement of JPEG applications. It should be noted that commonly researchers related to image classification methods are focused on object detection in images, i.e. pattern recognition, and only several investigations have been carried out to detect an effect of the JPEG algorithm to image classification. This effect on pattern recognition task was investigated in [7].

The paper [3] introduced an adaptive JPEG algorithm, which allows to compress images more efficiently taking into account their content. Smother images are compressed stronger, and images with many edges – weaker. MSE, SSIM, and Image Fidelity (IF) [15] as image quality assessment metrics were considered. The authors stated [3] that the metrics fail in the cases of images with high detail level, therefore with a view to detect image quality thresholds they used Laplace edge detection filter. However, the JPEG algorithm must be processed several times, and each time the compressed image has to be compared with the original one.

In the paper [1], a classification problem was solved in order to determine a relation between the extracted features of the human visual system (HVS) and the mean opinion score (MOS) by a growing and pruning radial-basis function (GAP-RBF) network. The authors aimed to create a metric that would closely correlate with human perception without needing the reference image. The following features were extracted: edge amplitude and length, background activity and luminance [6]. The edges were obtained using Prewitt edge operators [8]. The smallest error of the MOS prediction has been obtained by GAP-RBF network compared to other metrics investigated.

In the paper [2], a system that optimizes QF and scaling parameter of an image processed by the JPEG algorithm. The authors aimed to predict the optimal set of parameters for given maximum relative file size and viewing condition. However, the predictor does not predict resulting file size nor quality, but it predicts the transcoding parameters that maximize quality under the constraint of file size. Some improvements were introduced in [10] by the usage of k-means-based prediction of the transcoded JPEG file size and the SSIM index.

## 3   Proposed Classification-Based Compression Approach

In this paper, we propose an approach for the prediction and automatic selection of QF value when compressing images by the JPEG algorithm while maintaining minimal required quality level. Images are classified into two classes according to their quality after compression. Firstly, the proper metric for image quality assessment must be identify. In [3], ten different objective image quality metrics were compared, and the dependence of QF on the metrics values is evaluated. It was obtained that the values of the some metrics correlate, and only most distinctive ones should be used to get the generalized information on image quality. The least correlated metrics are SSIM and PSNR, therefore we address to both these metrics when evaluating the image quality.

Secondly, a minimal image quality level must be determined. Let's assume that images remain of high quality after JPEG compression, if SSIM $> q_1$ and PSNR $> q_2$. Otherwise, images are considered as low quality. The various values $q_1$ and $q_2$ can be chosen depending on image content and purpose of usage. Thus, a classification problem is to classify images into two classes with high or low quality after compressing images by the JPEG algorithm. The process of the proposed classification-based image compression approach is shown in Fig. 1 and detailed below. It should be noted that the training process is performed once, and the compression – on demand.

Let an image is described by $d$-dimensional vector $S = (s_1, s_2, \ldots, s_d)$, and $Y_1, Y_2, \ldots, Y_l$ are class labels, where $l$ is the number of classes. A general classification problem is solved by the following steps:

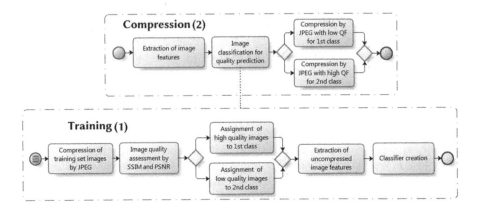

**Fig. 1.** Process of the proposed classification-based image compression approach

1. Considering to the vector $S$, the features $x_1, x_2, \ldots, x_n$, describing the image, are determined and the $n$-dimensional vector $X_i = (x_{i1}, x_{i2}, \ldots, x_{in})$ is formed, $n \leq d$. A set of the vectors $X_1, X_2, \ldots, X_m$ corresponds to a set of the images, where $m$ is the number of images. Each vector is assigned to one of classes $Y_j, j \in \{1, \ldots, l\}$.
2. The classifier is created (trained) using the vectors $X_1, X_2, \ldots, X_m$ corresponding to the images.
3. Considering to the classifier, an image of unknown class is assigned to one of known classes $Y_j, j \in \{1, \ldots, l\}$.

Here we select features describing images taking into consideration suggestions in the literature and results of our previous experimental investigations that have been carried out with a view to improve classification accuracy. The original image is represented by the vector $X_i = (x_{i1}, x_{i2}, \ldots, x_{i9}) = (Q_i, MR_i, MG_i, MB_i, SR_i, SG_i, SB_i, AB_i, AS_i)$. Here $Q_i$ shows physical memory used to save a pixel, i.e. it is a ratio between an image width multiplied by its height in pixels and a file size. The other features are obtained from images processed by Prewitt edge detection filter with high level for each color component in RGB (a high level results thick edges with low contrast and many colors in dark areas) (see Fig. 2). $MR_i, MG_i, MB_i$ are the means and $SR_i, SG_i, SB_i$ are the standard deviations of pixel values in each RGB component. $AB_i, AS_i$ are the number of large and small similar color areas in images processed by global image thresholding [9] (see Fig. 3).

## 4    Experimental Investigation

A digital image database SUN2012 [18] is used in this experimental investigation with a view to illustrate efficiency of the proposed classification-based image compression approach. The database consists of 16873 different images. 405 images are selected randomly, where the dimensions are not smaller than

$1024 \times 768$ pixels. For assignment of the images to classes (see Fig. 1, Training (1)), they are compressed by the JPEG algorithm with a low value of $QF = 30$, and the quality of the compressed images are computed. The obtained SSIM values range from 0.802 to 0.999, PSNR – from 25.7 to 57.3. Then the images are assigned to one of two classes ($l = 2$) by the following way:

– A set of 200 images with the highest SSIM values and a set of 200 images with the highest PSNR are selected. The images from the intersection of these two sets are assigned to the 1st class (116 images). The quality of these images does not change significantly after compression.
– The remaining images are assigned to the 2nd class (289 images). The quality of these images changes after compression more than that of the 1st class.

Now we return to uncompressed images, and the features describing them are extracted (see Sect. 3). As a majority of the features are extracted from images processed by Prewitt edge detection filter, for illustration, subsets of the 1st and 2nd class images, processed by this filter, are depicted in Fig. 2. An example of the large and small similar color areas is also presented in Fig. 3. Matlab environment is used for compressing images by the JPEG algorithm, for processing them in order to extract the features, for creating a classifier, and for assigning uncompressed images to classes.

Various classification methods can be applied to classify images considering to their quality after compression. Investigations have shown that in the case of image classification, the proper features describing images influence classification results more than the classification method used. Thus, for simplicity, here we use the linear discriminant analysis (LDA) as a classification method. After 10-fold cross-validation, the general classification accuracy 0.85 is obtained. This evaluation is not informative, as the number of members of each class differs. Therefore, it is purposeful to explore the confusion matrix:

|  |  | Prediction outcome | |
|---|---|---|---|
|  |  | 1st class | 2nd class |
| Actual value | 1st class | 66 | 50 |
|  | 2nd class | 11 | 278 |

Though, the classifier predicts the 1st class poorly (the true class is predicted for 66 images and the false class – for 50 images), however, the classification accuracy for the 2nd class is equal to 0.96. The classification results are satisfactory, as we will apply the high value of QF exactly for the images of the 2nd class (see Fig. 1, Compression (2)), thus, the image quality will not suffer. It should be noted that, in the case of the false prediction of the 1st class, the images of the 1st class will be assigned to the 2nd class, thus, they will be compressed with the high value of QF and their quality will not suffer as well.

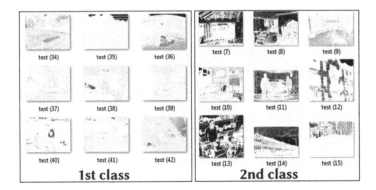

**Fig. 2.** Subsets of 1st and 2nd class images processed by Prewitt edge detection filter

**Fig. 3.** Image with the large and small similar color areas

The proposed classification-based approach is compared with the conventional JPEG algorithm in order to investigate storage space required for compressed images. The necessary condition is determined that the image quality after compression would satisfy such requirements: $q_1 = 0.91$ and $q_2 = 33$, thus, SSIM $> 0.91$ and PSNR $> 33$. These values ensure high image quality. In the conventional JPEG algorithm, high QF $= 95$ is set that the quality of all the images after compression satisfies these requirements. In our proposed image compression approach, different two values of QF are applied: high QF $= 95$ (the same as in the conventional JPEG algorithm) and low QF $= 30$. The low QF is selected that only images of the 1st class satisfy the quality requirements.

In each cross-validation round, a subset of images has been fixed. The images are assigned to either the 1st or 2nd class by the created classifier and compressed with corresponding QF value. The file sizes of the subsets of compressed images are presented in Table 1. In order to highlight a superiority of the proposed approach, the same images are also compressed by the conventional JPEG algorithm, and the results are also given in Table 1. The last row shows the total size of all the 405 image files. We see that, when using the proposed approach,

**Table 1.** Sizes of image files (MB) obtained after image compression

| Round no | Conventional JPEG | Proposed approach |
|---|---|---|
| 1 | 35.3 | 29.4 |
| 2 | 19.1 | 16.8 |
| 3 | 71.6 | 58.3 |
| 4 | 19.2 | 17.5 |
| 5 | 32.6 | 30.1 |
| 6 | 52.2 | 41.5 |
| 7 | 34.5 | 31.1 |
| 8 | 22.1 | 20.3 |
| 9 | 50.4 | 38.7 |
| 10 | 27.5 | 25.7 |
| **Total:** | **364.5** | **309.4** |

the required storage space is 15 % less than applying the conventional JPEG algorithm. Moreover, in the investigated set of images, there is no image that does not satisfy the quality requirements, i.e. the quality of the all 405 images exceeds the minimal SSIM and PSNR values.

## 5 Conclusions

In the paper, the classification-based image compression approach has been proposed and investigated. The approach consists of two main parts: (1) classifier training, and (2) image compression considering to classification results. The classifier is trained once, meanwhile image compression can be performed on demand. A superiority of the proposed approach is that the classification allows to predict and automatically select the proper value of QF for each image before compressing it by the JPEG algorithm. The approach enables to save storage space while maintaining sufficiently high image quality. The experimental investigation, where the proposed approach has been compared with the conventional JPEG algorithm, has shown that the approach allows to save 15 % of image storage space, when classifying only into two classes. The proposed approach is especially useful when compressing huge amount of images. With a view to save more image storage spaces, the further researches should include an improvement of the classification accuracy of each class as well as an investigation with a larger number of classes.

# References

1. Venkatesh Babu, R., Suresh, S.: GAP-RBF based NR image quality measurement for JPEG coded images. In: Kalra, P.K., Peleg, S. (eds.) ICVGIP 2006. LNCS, vol. 4338, pp. 718–727. Springer, Heidelberg (2006). doi:10.1007/11949619_64

2. Coulombe, S., Pigeon, S.: Low-complexity transcoding of JPEG images with near-optimal quality using a predictive quality factor and scaling parameters. IEEE Trans. Image Process. **19**(3), 712–721 (2010)

3. Forczmański, P., Mantiuk, R.: Adaptive and quality-aware storage of JPEG files in the web environment. In: Chmielewski, L.J., Kozera, R., Shin, B.-S., Wojciechowski, K. (eds.) ICCVG 2014. LNCS, vol. 8671, pp. 212–219. Springer, Heidelberg (2014). doi:10.1007/978-3-319-11331-9_26

4. Hore, A., Ziou, D.: Image quality metrics: PSNR vs. SSIM. In: 20th International Conference on Pattern Recognition (ICPR), pp. 2366–2369. IEEE (2010)

5. IJG: Independent JPEG Group (2015). http://www.ijg.org/

6. Karunasekera, S.A., Kingsbury, N.G.: A distortion measure for blocking artifacts in images based on human visual sensitivity. IEEE Trans. Image Process. **4**(6), 713–724 (1995)

7. Lau, W.L., Li, Z.L., Lam, K.K.: Effects of JPEG compression on image classification. Int. J. Remote Sens. **24**(7), 1535–1544 (2003)

8. Maini, R., Aggarwal, H.: Study and comparison of various image edge detection techniques. Int. J. Image Process. (IJIP) **3**(1), 1–11 (2009)

9. Otsu, N.: A threshold selection method from gray-level histograms. Automatica **11**(285–296), 23–27 (1975)

10. Pigeon, S., Coulombe, S.: K-means based prediction of transcoded JPEG file size and structural similarity. Int. J. Multimed. Data Eng. Manage. (IJMDEM) **3**(2), 41–57 (2012)

11. Ponomarenko, N., Ieremeiev, O., Lukin, V., Egiazarian, K., Jin, L., Astola, J., Vozel, B., Chehdi, K., Carli, M., Battisti, F., et al.: Color image database TID2013: peculiarities and preliminary results. In: 4th European Workshop on Visual Information Processing (EUVIP), pp. 106–111. IEEE (2013)

12. Salomon, D.: A Guide to Data Compression Methods. Springer, Heidelberg (2013)

13. Skodras, A., Christopoulos, C., Ebrahimi, T.: The JPEG2000 still image compression standard. IEEE Signal Process. Mag. **18**(5), 36–58 (2001)

14. Wallace, G.K.: The JPEG still picture compression standard. IEEE Trans. Consu. Electron. **38**(1), xviii–xxxiv (1992)

15. Wang, Z., Bovik, A.C.: Modern image quality assessment. Synth. Lect. Image Video Multimed. Process. **2**(1), 1–156 (2006)

16. Wang, Z., Bovik, A.C.: Mean squared error: love it or leave it? a new look at signal fidelity measures. IEEE Signal Process. Mag. **26**(1), 98–117 (2009)

17. Wang, Z., Bovik, A.C., Sheikh, H.R., Simoncelli, E.P.: Image quality assessment: from error visibility to structural similarity. IEEE Trans. Image Process. **13**(4), 600–612 (2004)

18. Xiao, J., Hays, J., Ehinger, K.A., Oliva, A., Torralba, A.: SUN database: large-scale scene recognition from abbey to zoo. In: 2010 IEEE Conference on Computer Vision and Pattern Recognition (CVPR), pp. 3485–3492. IEEE (2010)

# Image Despeckling Using Non-local Means with Diffusion Tensor

Mariusz Nieniewski$^{(\boxtimes)}$ and Paweł Zajączkowski

Department of Mathematics and Informatics, University of Lodz,
ul. Banacha 22, 90-238 Lodz, Poland
{mnieniew.pawel,zajaczkowski.pawel}@math.uni.lodz.pl

**Abstract.** This paper presents a novel modification of one of the varieties of the non-local means (NLM) algorithm for speckle reduction in images. This modification comes in the form of replacement of the structure tensor used in the NLM algorithm by the diffusion tensor. The diffusion tensor originally was used in the nonlinear coherent diffusion algorithm making possible intensification of the diffusion in the direction parallel to edges and inhibition in the direction perpendicular to edges. It is shown in this paper that using the diffusion tensor in calculating the weights for the NLM leads to an improvement of the quality of despeckled images. The NLM algorithm has a tendency to smooth the image in such a way that the despeckled image is covered by relatively flat areas typical for mosaic images. This tendency is undesirable since flat areas form visible contours that are not related to the object visualized. The superiority of the new despeckling filter is confirmed by examples of filtering the ultrasound (US) images as well as by image quality measures.

**Keywords:** Non-local means · Diffusion tensor · Image despeckling · Ultrasound images

## 1 Introduction

The NLM is one of the image despeckling methods and was originally presented in [2]. Subsequently several important modifications of this method were developed. In particular, in [3] a Bayesian framework is used to derive a NLM algorithm adapted to the US image noise model. In [6] the anisotropic NLM algorithm is proposed for general images. In [9] the NLM filter is presented which employs the structural similarity index instead of Euclidean distance in the calculation of the weights used by the NLM filter. In [10] the brightness distribution vectors used for calculating weights are projected on a lower dimension space using PCA. In [13] Wu presents a modification of the classic NLM in which the Euclidean distance between the brightness vectors used in the calculation of the weights is replaced by the structure tensor taking into account specifically the possibility of occurrence of edges in the image. The aim of the current paper is a further improvement of the Wu's method [13] by replacing the structure tensor by

© Springer International Publishing AG 2017
R.S. Choraś (ed.), *Image Processing and Communications Challenges 8*,
Advances in Intelligent Systems and Computing 525, DOI 10.1007/978-3-319-47274-4_5

the diffusion tensor. Taking the mechanism of the diffusion tensor from another despeckling method, that is a nonlinear coherent diffusion [8], and applying it to the NLM one hopes to improve the effectiveness of despeckling because the nonlinear relation between the structure tensor and the diffusion tensor has an advantageous effect on the diffusion. The subsequent sections show that this in fact is true and that replacement of the structure tensor by the diffusion tensor improves the quality of the despeckled images.

## 2  Theoretical Considerations of NLM

### 2.1  Classical NLM (NLMC)

The classical NLM filter averages pixel intensities weighted by the similarity of pixel gray levels in a certain neighborhood. Suppose we filter the noisy input image $I$ by means of NLM and obtain the output image $O$. The NLM method is based on the following Eq. [2]

$$O(p) = \sum_{q \in \Omega} w(p,q) I(q) \tag{1}$$

where $O(p)$ is the calculated brightness at pixel $p$, and $\Omega$ is some area in the image. The $I(q)$ is the unfiltered brightness of a pixel $q$ in the neighborhood of the pixel $p$ in the input image $I$, and the weight $w(p,q)$ is calculated using the Gaussian weighting function

$$w(p,q) = \frac{1}{Z(p)} \exp\left\{ -\frac{\|B(p) - B(q)\|^2}{h^2} \right\} \tag{2}$$

In this equation $h$ is a filtering parameter, and $B(p)$ and $B(q)$ denote the vectors containing the brightness of neighboring pixels centered around $p$ and $q$, respectively, hence $\|B(p) - B(q)\|$ is the Euclidean distance between the two brightness vectors $B(p)$ and $B(q)$.

$Z(p)$ in Eq. (2) is a normalizing constant calculated by means of the equation

$$Z(p) = \sum_{q \in \Omega} \exp\left\{ -\frac{\|B(p) - B(q)\|^2}{h^2} \right\} \tag{3}$$

In summary, the value $O(p)$ is calculated as a weighted mean of pixels in a certain area $\Omega$. The weights are chosen so that the greater weight is assigned to pixels which have brightness distribution around them similar to that of the pixel under consideration.

### 2.2  NLM with Structure Tensor (NLMST)

The use of the structure tensor in the NLM was introduced by [13]. In this paper we assume the formulation of the structure tensor similar to [8]. First the directional gradients $I_x(p), I_y(p)$ in the $x$ and $y$ directions are calculated for the input image $I(p)$. Subsequently the components of the structure tensor $J(p)$ are obtained as a convolution of $I_x^2(p)$, $I_x(p)I_y(p)$, $I_y^2(p)$ with a chosen square Gaussian mask of side $s_1$ and standard deviation $\sigma$ [8]. Subsequently, the structure tensor is written as

$$J(p) = \begin{bmatrix} J_{11}(p) & J_{12}(p) \\ J_{21}(p) & J_{22}(p) \end{bmatrix} \equiv \begin{bmatrix} a & b \\ b & d \end{bmatrix} \quad (4)$$

with $J_{21}(p) = J_{12}(p)$. Then the distance between the tensors has to be used rather than the distance between vectors. Following [13] we use the Riemannian metric called Log-Euclidean metric. The advantages of using this metric for tensor comparisons are considered both in [13] and several papers cited therein. In accordance with these considerations we have the following metric $||J(p) - J(q)||$ for the definition of the distance between tensors for pixels $p$ and $q$

$$||J(p) - J(q)|| = \sqrt{trace\{[log(J(p)) - log(J(q))]^2\}} \quad (5)$$

The weighting function is now

$$w(p,q) = \frac{1}{Z(p)} \exp\left\{ -\frac{||B(p) - B(q)||^2 + \beta||J(p) - J(q)||^2}{h^2} \right\} \quad (6)$$

The parameter $\beta$ in the above equation represents the strength of the influence of the structure tensor on the calculated weight. The normalizing constant $Z(p)$ is now defined as

$$Z(p) = \sum_{q \in \Omega} \exp\left\{ -\frac{||B(p) - B(q)||^2 + \beta||J(p) - J(q)||^2}{h^2} \right\} \quad (7)$$

It is worth noting that for $\beta = 0$ we obtain the classical NLM.

## 2.3   NLM with Diffusion Tensor (NLMDT)

The diffusion tensor is obtained from the structure tensor Eq. (4) in such a way that we keep the structure tensor's eigenvectors and modify its eigenvalues [8]. The eigenvalues of the structure tensor are

$$\mu_{1,2} = \frac{1}{2}[a + d \pm \sqrt{a^2 + 4b^2 - 2ad + d^2}] \quad (8)$$

where $\mu_1 > \mu_2$ and the eigenvectors are

$$\begin{bmatrix} \{a - d + \sqrt{a^2 + 4b^2 - 2ad + d^2}\}/2 \\ b \end{bmatrix} = \begin{bmatrix} w_a \\ w_b \end{bmatrix} \quad (9)$$

$$\begin{bmatrix} \{a - d - \sqrt{a^2 + 4b^2 - 2ad + d^2}\}/2 \\ b \end{bmatrix} = \begin{bmatrix} w_d \\ w_b \end{bmatrix} \quad (10)$$

After normalization the eigenvectors are rewritten as $[v_a \; v_b]^T$, $[v_c \; v_d]^T$ with components $v_a, \ldots, v_d$ used in place of $w_a, \ldots, w_d$.

Then the eigenvalues $\lambda_1, \lambda_2$ of the diffusion tensor are defined by the equations

$$\lambda_1 = \begin{cases} \alpha(1 - \frac{(\mu_1 - \mu_2)^2}{s^2}) & \text{if } (\mu_1 - \mu_2)^2 \leq s^2 \\ 0 & \text{otherwise} \end{cases} \tag{11}$$

$$\lambda_2 = \alpha$$

with parameters $\alpha$ and $s$ controlling the diffusion tensor similarly to what goes on in the case of nonlinear coherent diffusion. More specifically, the diffusion tensor is defined as a product of three matrices

$$D(p) = \begin{bmatrix} v_a & v_c \\ v_b & v_d \end{bmatrix} \begin{bmatrix} \lambda_1 & 0 \\ 0 & \lambda_2 \end{bmatrix} \begin{bmatrix} v_a & v_b \\ v_c & v_d \end{bmatrix} \equiv \begin{bmatrix} a_d & b_d \\ c_d & d_d \end{bmatrix} \tag{12}$$

According to Eq. (11) the eigenvalue $\lambda_1$ is a nonlinear function of the difference of $\mu_1$ and $\mu_2$. If $\mu_1 = \mu_2$ then $\lambda_1 = \lambda_2 = \alpha$. Otherwise $\lambda_1$ decreases. In nonlinear coherent diffusion the images are smoothed in such a way that the diffusion is carried out along the edges and is extinguished in the directional perpendicular to the edges. In our case changes in diffusion intensity are replaced by modifications in the calculation of the weights.

Having defined the diffusion tensor we can substitute the diffusion tensor $D(p)$ in place of the structure tensor $J(p)$ in Eqs. (5)–(7) and perform all the calculations in an analogous manner. It is worth noting that the NLM filtering may be carried out iteratively [1].

## 3    Experimental Results

The parameters assumed in our experiments are as follows. The filtering parameter in Eqs. (2) and (6) is $h = 10$ and is based on [2,13]. The size of the window for calculating the vectors $B$ and $J$ in Eqs. (2) and (6) is $5 \times 5$ pixels [2]. The size of the area $|\Omega|$ is $11 \times 11$ pixels in accordance with the suggestion in [2]. The Gaussian mask used for filtering prior to the formulation of the structure tensor is a square of side $s_1 = 9$ and standard deviation $\sigma = 15$, similarly to [8]. The parameters used in Eq. (11) for the diffusion tensor calculations are $s = 120$ and $\alpha = 1$ and were taken from [8]. The strength parameter in Eq. (6) is $\beta = 20$ unless otherwise stated below.

Two examples of unfiltered US images are shown in Fig. 1, and their despeckled counterparts, shown in Figs. 2 and 3, are evaluated using the image quality measures specified in [4,7]. The meaning of measures' acronyms is as follows: MSE (mean square error), RMSE (root mean square error), ERR3 (norm of dissimilarity between the original and despeckled images calculated as Minkowski norm with the parameter equal to 3), ERR4 (Minkowski norm with the parameter equal to 4), GAE (geometric average error), SNR (signal-to-noise ratio), PSNR (peak signal-to-noise ratio), Q (universal image quality index), SSIN (structural similarity index; denoted as SSIM in [9]).

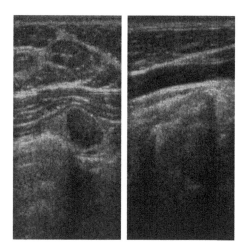

**Fig. 1.** Left: original image of a breast. Right: original image of a carotid artery, adapted from [7]

**Table 1.** Image quality measures for breast images in Fig. 2 displayed in successive rows after 1 and 4 iterations

| Measure | NLMC | NLMST | NLMDT | Measure | NLMC | NLMST | NLMDT |
|---------|------|-------|-------|---------|------|-------|-------|
| MSE | 60.203 | 59.892 | 19.011 | SNR | 24.675 | 24.697 | 29.695 |
|     | 196.104 | 190.992 | 86.365 |     | 19.510 | 19.627 | 23.105 |
| RMSE | 7.759 | 7.739 | 4.360 | PSNR | 30.335 | 30.357 | 35.341 |
|      | 14.004 | 13.820 | 9.293 |      | 25.206 | 25.321 | 28.767 |
| ERR3 | 9.173 | 9.150 | 6.346 | Q | 0.814 | 0.815 | 0.943 |
|      | 16.550 | 16.331 | 11.521 |   | 0.393 | 0.405 | 0.717 |
| ERR4 | 10.462 | 10.437 | 8.063 | SSIN | 0.820 | 0.821 | 0.945 |
|      | 18.866 | 18.615 | 13.515 |      | 0.415 | 0.426 | 0.728 |
| GAE (1–4) | 0 | 0 | 0 |  |  |  |  |

All the calculations are coded in Matlab with the code for NLMC originating from [5]. Despeckling the breast image from Fig. 1 is illustrated in Fig. 2 for the three NLM filters under consideration and for 1 − 4 iterations. Similarly, despeckling the carotid artery image from Fig. 1, is depicted in Fig. 3. Analyzing the above images one comes to the conclusion that the NLMLST and NLMC give very similar results. The filtered images, particularly for 3 and 4 iterations are to a large extent despeckled, but at the same time exhibit flat areas with very distinct contours. The NLMDT images are devoid of such areas and seem more natural.

Most of the above quality measures have a fairly obvious interpretation. The most sophisticated are the universal quality index Q [11] and structural similarity index SSIN originally developed in [12]. Both Q and SSIN are functions of the

**Table 2.** Image quality measures for carotid artery images in Fig. 3 displayed in successive rows after 1 and 4 iterations

| Measure | NLMC | NLMST | NLMDT | Measure | NLMC | NLMST | NLMDT |
|---------|------|-------|-------|---------|------|-------|-------|
| MSE | 58.082 | 57.810 | 18.260 | SNR | 22.806 | 22.826 | 27.856 |
|     | 158.530 | 155.356 | 87.417 |     | 18.406 | 18.497 | 21.025 |
| RMSE | 7.621 | 7.603 | 4.273 | PSNR | 30.490 | 30.511 | 35.516 |
|      | 12.591 | 12.464 | 9.350 |      | 26.130 | 26.218 | 28.715 |
| ERR3 | 8.929 | 8.908 | 6.108 | Q | 0.693 | 0.694 | 0.915 |
|      | 14.962 | 14.806 | 11.348 |   | 0.285 | 0.293 | 0.511 |
| ERR4 | 10.096 | 10.073 | 7.693 | SSIN | 0.713 | 0.714 | 0.919 |
|      | 17.160 | 16.979 | 13.140 |      | 0.325 | 0.333 | 0.544 |
| GAE (1–4) | 0 | 0 | 0 | | | | |

following parameters determined for the original image and despeckled image: mean value, standard deviation, and the covariance between the original and despeckled image. In fact, the SSIN is a generalization of Q. Both Q and SSIN are contained in the range $[-1, 1]$ and their higher values indicate better agreement of the despeckled image with the original image.

The image quality measures for Figs. 2 and 3 are given in Tables 1 and 2 respectively. Comparing Q as well as SSIN in Table 1 for a given iteration number we conclude that their best values are obtained for the NLMDT. Similar observations can be made with respect to the other quality measures in Table 1 as well as Table 2.

Figure 4 illustrates the dependence of the image quality measures Q and SSIN on the strength parameter $\beta$ for 1 and 4 iterations. It can be seen from this figure that for $5 \leq \beta \leq 100$ the image quality is almost independent of $\beta$. One noteworthy point in the diagrams of Fig. 4 is that corresponding to $\beta = 0$, which should be used as a reference for all the other points. It is evident from Fig. 4 that the improvement for the NLMST in comparison with NLM is marginal whereas the improvement for NLMDT is much more pronounced. Of course, Q and SSIN go down when the number of iterations is increased. This is the result of the fact that the original image is the noisy one and not the ideal unspeckled image which does not exist. A set of 10 US images originally used in [7] and representing a carotid artery, elbow, thyroid, abdomen, breast, ligament, muscle, and a vein were also tested for the purposes of the current paper. The mean values of the image quality measures obtained for these images are specified in Table 3. The conclusions drawn from the inspection of this table are similar to the ones given above.

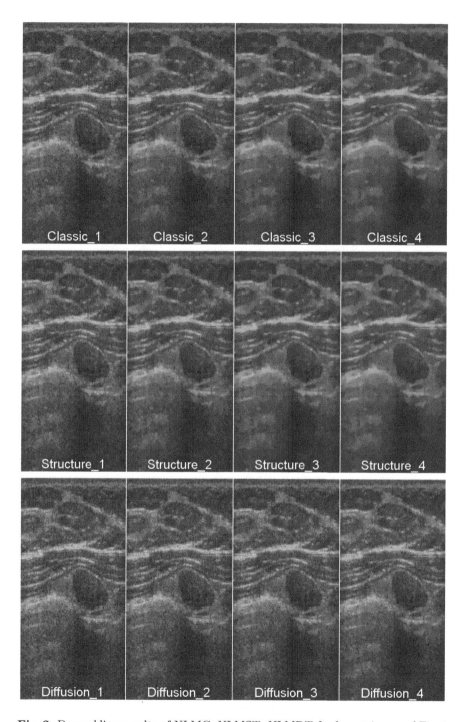

**Fig. 2.** Despeckling results of NLMC, NLMST, NLMDT for breast image of Fig. 1

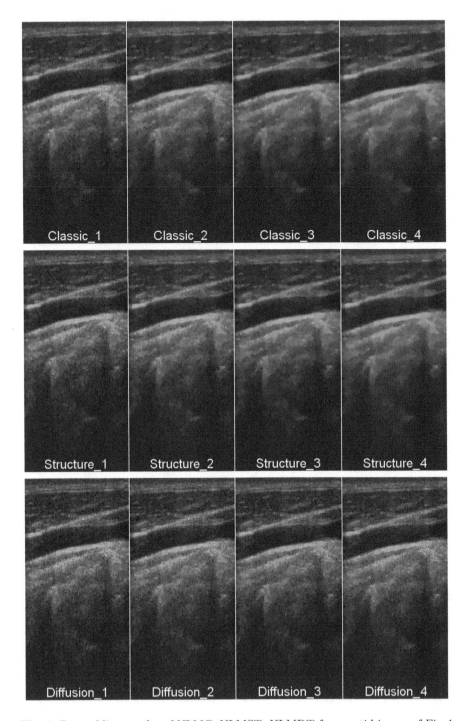

**Fig. 3.** Despeckling results of NLMC, NLMST, NLMDT for carotid image of Fig. 1

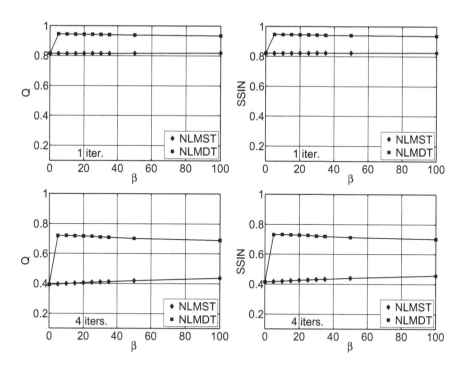

**Fig. 4.** The universal quality index Q and structural similarity index SSIN for the breast image of Fig. 1 calculated as a function of the strength parameter $\beta$ for 1 and 4 iterations of NLMST and NLMDT

**Table 3.** Mean values of image quality measures for a set of 10 images, displayed in successive rows after 1 and 4 iterations

| Measure | NLMC | NLMST | NLMDT | Measure | NLMC | NLMST | NLMDT |
|---|---|---|---|---|---|---|---|
| MSE | 58.475 | 58.163 | 18.912 | SNR | 23.300 | 23.324 | 28.227 |
|  | 179.489 | 175.118 | 84.937 |  | 18.401 | 18.511 | 21.676 |
| RMSE | 7.646 | 7.626 | 4.349 | PSNR | 30.462 | 30.485 | 35.364 |
|  | 13.380 | 13.217 | 9.215 |  | 25.612 | 25.718 | 28.841 |
| ERR3 | 8.998 | 8.975 | 6.278 | Q | 0.762 | 0.763 | 0.927 |
|  | 15.845 | 15.649 | 11.331 |  | 0.354 | 0.364 | 0.638 |
| ERR4 | 10.218 | 10.193 | 7.948 | SSIN | 0.775 | 0.776 | 0.930 |
|  | 18.102 | 17.877 | 13.228 |  | 0.383 | 0.393 | 0.659 |
| GAE (1–4) | 0 | 0 | 0 |  |  |  |  |

# 4 Conclusions

Replacing the structure tensor by the diffusion tensor in calculating the weights used in the NLM leads to the improvement of the despeckled images. The superiority of the proposed NLMDT method over the NLMST is confirmed by both visual inspection of images as well as calculation of several image quality measures. One possible variety of the described research would be inclusion of the structure tensor and the diffusion tensor each with its own strength parameter. The presented NLMDT method of despeckling is quite general and can be used for any kind of images, not necessarily US medical images. For example, Wu [13] tested the NLMST on natural images artificially contaminated with noise.

**Acknowledgments.** The authors would like to express their gratitude to Prof. Andrzej Nowicki, Institute of Fundamental Technological Research, Warsaw for providing the necessary images for this research.

# References

1. Brox, T., Cremers, D.: Iterated nonlocal means for texture restoration. In: Sgallari, F., Murli, A., Paragios, N. (eds.) SSVM 2007. LNCS, vol. 4485, pp. 13–24. Springer, Heidelberg (2007). doi:10.1007/978-3-540-72823-8_2

2. Buades, A., Coll, B., Morel, J.M.: A review of image denoising algorithm, with a new one. Multiscale Model. Simul. **4**(2), 490–530 (2005)

3. Coupé, P., Hellier, P., Kervrann, C., Barillot, C.: Nonlocal means-based speckle filtering for ultrasound images. IEEE Trans. Image Process. **18**(10), 2221–2229 (2009)

4. Loizou, C.P., Pattichis, C.S.: Despeckle Filtering for Ultrasound Imaging and Video. Morgan & Claypool, San Rafael (2015)

5. Manjon-Herrera, J.V.: Non-local means filter. http://www.mathworks.com/matlabcentral/fileexchange/13176-non-local-means-filter. Last accessed Jun 2016

6. Maleki, A., Narayan, M., Baraniuk, R.G.: Anisotropic nonlocal means denoising. App. Comput. Harmon. Anal. **35**, 452–482 (2011)

7. Nieniewski, M.: Enhancement of despeckled ultrasound images by forward-backward diffusion. In: Chmielewski, L.J., Kozera, R., Shin, B.-S., Wojciechowski, K. (eds.) ICCVG 2014. LNCS, vol. 8671, pp. 454–461. Springer, Heidelberg (2014). doi:10.1007/978-3-319-11331-9_54

8. Nieniewski, M., Zajączkowski, P.: Real-time speckle reduction in ultrasound images by means of nonlinear coherent diffusion using GPU. In: Chmielewski, L.J., Kozera, R., Shin, B.-S., Wojciechowski, K. (eds.) ICCVG 2014. LNCS, vol. 8671, pp. 462–469. Springer, Heidelberg (2014). doi:10.1007/978-3-319-11331-9_55

9. Rehman, A., Wang, Z.: SSIM-based non-local means image denoising. In: 18th IEEE International Conference on Image Processing, pp. 217–220 (2011)

10. Tasdizen, T.: Principal neighborhood dictionaries for non local means image denoising. IEEE Trans. Image Process. **18**(12), 2649–2660 (2009)

11. Wang, Z., Bovik, A.: A universal quality index. IEEE Sig. Process. Lett. **9**(3), 81–84 (2002)

12. Wang, Z., Bovik, A., Sheikh, H., Simoncelli, E.: Image quality assessment: from error measurement to structural similarity. IEEE Trans. Image Process. **13**(4), 600–613 (2004)

13. Wu, X., Xie, M., Wu, W., Zhou, J.: Nonlocal mean image denoising using anisotropic structure tensor. Adv. Opt. Technol. **2013** (2013), http://dx.doi.org/10.1155/2013/794728. Article ID 794728. Hindawi, Cairo

# Face Recognition with 3D Face Asymmetry

Janusz Bobulski[(✉)]

Institute of Computer and Information Sciences,
Czestochowa University of Technology, Dabrowskiego 73,
42-200 Czestochowa, Poland
januszb@icis.pcz.pl

**Abstract.** Using of 3D images for the identification was in a field of the interest of many researchers which developed a few methods offering good results. However, there are few techniques exploiting the 3D asymmetry amongst these methods. We propose fast algorithm for rough extraction face asymmetry that is used to 3D face recognition with hidden Markov models. This paper presents conception of fast method for determine 3D face asymmetry. The research results indicate that face recognition with 3D face asymmetry may be used in biometrics systems.

**Keywords:** Face asymmetry · Hidden Markov models · Face recognition · Identity verification

## 1 Introduction

Biometrics systems use individual and unique biological features of person for user identification. The most popular features are: fingerprint, iris, voice, palm print, face image et al. Most of them are not accepted by users, because they feel under surveillance or as criminals. Others, in turn, are characterized by problems with the acquisition of biometric pattern and require closeness to the reader. Among the biometric methods popular technique is to identify people on the basis of the face image, the advantage is the ease of obtaining a biometric pattern. Low prices of cameras have caused their commonness and they are everywhere. Moreover, the quality of the images captured from modern cameras are so good that they may be used to retrieve biometric patterns, and then for identification. The advantage of the identification with the face image is the ease acquiring pattern and a high acceptance level of this method by users. There are many works on 2D face recognition [20], and made great progress in this field. Among these works there are also techniques that use the asymmetry of the face, and the efficiency of this technique is confirmed in articles [9,10,13,19].

With the development of 3D technology appeared methods of 3D face recognition. In last years, some of the new face recognition strategies tend to overcome face recognition problem from a 3D perspective. The 3D data points proper to the surface of the face give us other kind of information for recognition, and solve the problem of pose and lighting variations in case of 2D data. However,

© Springer International Publishing AG 2017
R.S. Choraś (ed.), *Image Processing and Communications Challenges 8*,
Advances in Intelligent Systems and Computing 525, DOI 10.1007/978-3-319-47274-4_6

3D images have their own problems, e.g. normalization, devices for acquiring faces, time and cost of faces getting [12]. In the literature, we may find a lot of useful reviews of 3D face recognition problem such as [1].

Many works are dedicated to the 3D face recognition problem. There is the method presented by Riccio et al. [16] among them, that uses predefined key-points. These points are used to indicate the several geometric invariants on the basis of which is made identification. Other method, Rama et al. present in arti-cle [15]. They propose Partial Principle Component Analysis ($P^2CA$) for feature extraction and dimensionality reduction by projection 3D data into cylindrical coordinate. In [2], researchers use the iterative closest point (ICP) to adjust the 3D surface points of a face and then realize the recognition based on the mini-mum distance between the two faces. These methods have high recognition rate, but their main problem is speed and computational complexity.

Using of 3D images for the identification was in a field of the interest of many researchers which developed a few methods offering good results [4]. However, there are few techniques exploiting the 3D asymmetry amongst these methods. The reason for this is, among others, the problem of obtaining 3D images. The cost 3D camera is still higher than traditional camera and therefore their popu-larity and prevalence is lower. The second major problem in the processing of 3D images is their quality. Imperfection devices for image acquisition cause errors in the measurements and data discontinuity, that is a significant problem in the further processing of the data. At the present moment, however, we need to use the data in the quality of such is, and try to eliminate the disadvantages of these data and develop more effective methods of asymmetry measurement and face recognition based on asymmetry.

Few papers in the literature are dedicated to the 3D asymmetry face recogni-tion task so far. Huang et al. [7] propose method based on Local Binary Pattern (LBP). Their approach splits the face recognition task into two steps: (1) a matching step respectively processed in 2D/2D; (2) 3D/2D a fusion step com-bining two matching scores. Canonical Correlation Analysis (CCA) is applied in method propose by Yang et al. [18]. They apply CCA to learn the mapping between the 2D face image and 3D face data, and only 3d data is used for enrolment and recognition.

This article presents face recognition method based on 3D face asymmetry. We propose fast algorithm for rough extraction face asymmetry that is used to 3D face recognition with hidden Markov models (HMM) [3].

## 2   Proposed Method

### 2.1   Preprocessing

The pre-processing procedure of the system consists of the following steps:

- selection of face area.
- scaling image;
- rotation;

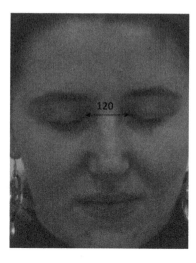

**Fig. 1.** Result of pre-processing procedure - scaling

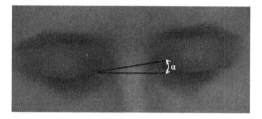

**Fig. 2.** Please write your figure caption here

$$\alpha = atan(y_2 - y_1, x_2 - x_1) \tag{1}$$

The main area of the face selected and rejected areas that contain little useful information on the outskirts of face. The selection of face area made based on keypoints [6], and the coordinates of these points are obtained from database. Based on inner corners of the eyes, the face image is scaled so that the distance between them was equal to 120 pixels (Fig. 1). Next, the angle of rotation is calculated from the mentioned coordinates (Eq. 1), and face image is rotated by an angle $\alpha$ (Fig. 2). This operation is aimed at establishing the identical position for all faces.

## 2.2 Measurement of the Asymmetry

There are many methods to found vertical line of face asymmetry. Ostwald et al. [14] propose a definition of the line asymmetry so that the differences between the face and its mirror reflection are as low as possible. Other method is proposed by Kurach et al. [11]. They propose to appoint line asymmetry in such a way

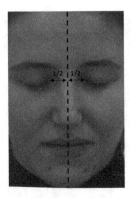

**Fig. 3.** Line of asymmetry

that the differences between the left and right part of the face are as small as possible. We propose simple and fast method of designate the line of asymmetry. The coordinates of keypoints points obtained from database exploit to find the centre of line connecting the inner corners of the eyes. Thus obtained value is used to determine the x-coordinate defining the lines of facial asymmetry (Fig. 3).

In this way we are dividing the face into the right and left part. Through the mirror vertically they are rising from these parts right face (RF) and left face (LF). From z-coordinate of these two elements and the normal face (NF) the measurement of the asymmetry is being made. In this way, the three metrics are formed that are differences between the RF, LF and NF (Eqs. 2–4) (Fig. 4).

$$LN = |LF - NF| \tag{2}$$

$$RN = |RF - NF| \tag{3}$$

$$LR = |LF - RF| \tag{4}$$

### 2.3  Recognition System

We have two basic tasks in face recognition application: learning and testing. In case of HMM [17], first task is made with Baum-Welch algorithm, that is based on the forward-backward algorithm. Second task may be made in some ways, but we chose forward algorithm.
Forward Algorithm [8]:
Define forward variable $\alpha_t(i)$ as:

$$\alpha_t(i) = P(o_1, o_2, , o_t, q_t = i | \lambda) \tag{5}$$

$\alpha_t(i)$ is the probability of observing the partial sequence $(o_1, o_2, , o_t)$ such that the the state $q_t$ is $i$

$$\alpha_{t+1}(i) = \left[ \sum_{i=1}^{N} \alpha_t(i) a_{ij} \right] b_j(o_{t+1}) \tag{6}$$

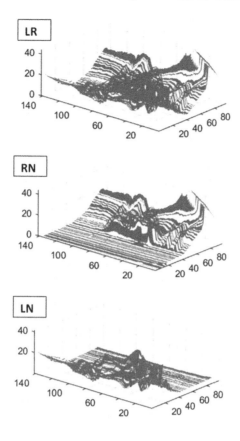

**Fig. 4.** Results of the measurement of the face asymmetry

Backward Algorithm [8]:

Define backward variable $\beta_t(i)$ as:

$$\beta_t(i) = P(o_{t+1}, o_{t+2}, , o_T, q_t = i | \lambda) \tag{7}$$

$\beta_t(i)$ is the probability of observing the partial sequence $(o_1, o_2, , o_t)$ such that the the state $q_t$ is $i$

$$\beta_t(i) = \sum_{i=1}^{N} a_{ij} b_j (o_{t+1} \beta_{t+1}(j)), \tag{8}$$

$1 \leq i \leq N, t = T - 1, ..., 1.$

Baum-Welch Algorithm [8]:

Define $\xi(i, j)$ as the probability of being in state $i$ at time $t$ and in state $j$ at time $t + 1$

$$\xi(i, j) = \frac{\alpha_t(i) a_{ij} b_j (o_{t+1}) \beta_{t+1}(j)}{P(O|\lambda)} = \frac{\alpha_t(i) a_{ij} b_j (o_{t+1}) \beta_{t+1}(j)}{\sum_{i=1}^{N} \sum_{j=1}^{N} \alpha_t(i) a_{ij} b_j (o_{t+1}) \beta_{t+1}(j)} \tag{9}$$

Define $\gamma(i)$ as the probability of being in state $i$ at time $t$, given observation sequence.

$$\gamma_t(i) = \sum_{j=1}^{N} \xi_t(i,j) \tag{10}$$

Update rules:

- $\bar{\pi}_i$ = expected frequency in state $i$ at time $(t = 1) = \gamma_1(i)$
- $\bar{a}_{ij}$ = (expected number of transition from state $i$ to state $j$)/(expected number of transitions from state $i$:

$$\bar{a}_{ij} = \frac{\sum_t \xi_t(i,j)}{\sum_t \gamma_t(i)} \tag{11}$$

- $\bar{b}_j(k)$ = (expected number of times in state $j$ and oserving symbol $k$)/(expected number of times in state $j$:

$$\bar{b}_j(k) = \frac{\sum_{t,o_t=k} \gamma_t(j)}{\sum_t \gamma_t(j)} \tag{12}$$

## 3  Experiments

In experiments we used the image database UMB-DB. The University of Milano Bicocca 3D face database is a collection of multimodal (3D + 2D colour images) facial acquisitions. The database is available to universities and research centres interested in face detection or face recognition. They recorded 1473 images of 143

**Table 1.** Results of experiments

| Type of asymetry | No. of test set | Recognition rate |
|---|---|---|
| LN | 1 | 58 % |
| LN | 2 | 62 % |
| LN | 3 | 60 % |
| Average | | 60 % |
| RN | 1 | 58 % |
| RN | 2 | 60 % |
| RN | 3 | 62 % |
| Average | | 60 % |
| LR | 1 | 68 % |
| LR | 2 | 70 % |
| LR | 3 | 72 % |
| Average | | 70 % |

**Table 2.** Comparison to other methods

| Method | Recognition rate |
|--------|------------------|
| LBP    | 82 %             |
| CCA    | 68 %             |
| Our    | 70 %             |

subjects (98 male, 45 female). The images show the faces in variable condition, lighting, rotation and size [5]. We chose three datasets, each consist of 50 persons in order to verify the method, and for each individual chose two images for learning and two for testing. The HMM implemented with parameters N = 10, O = 20. Table 1 presents the results of experiments.

## 4   Conclusion

This paper presented conception of fast and rough method for determines 3D face asymmetry. Presented method allows for faster 3D face processing and recognition because they do not use complex calculation for features extraction. The obtained results are satisfactory in comparison to other method and proposed method may be the alternative solution to the others (Table 2). Experiments confirmed the validity of the concept of 3D face asymmetry, and it is a faster method in comparison to another. The research results indicate that face recognition with 3D face asymmetry may be used in biometrics systems.

## References

1. Abate, A., Nappi, M., Riccio, D., Sabatino, G.: 2D and 3D face recognition: a survey. Pattern Recogn. Lett. **28**(14), 1885–1906 (2007)
2. Beumier, C., Acheroy, M.: Automatic 3D face authentication. Image Vis. Comput. **18**(4), 315–321 (2000)
3. Bobulski, J.: 2DHMM-based face recognition method. In: Choraś, R.S. (ed.) Image Processing and Communications Challenges 7. AISC, vol. 389, pp. 11–18. Springer, Heidelberg (2016). doi:10.1007/978-3-319-23814-2_2
4. Bowyer, K., Chang, K., Flynn, P.: A survey of approaches and challenges in 3D and multi-modal 3D+2D face recognition. Comput. Vis. Image Underst. **101**, 1–15 (2006)
5. Colombo, A., Cusano, C., Schettini, R.: UMB-DB: a database of partially occluded 3D Faces. In: Proceedings of ICCV 2011 Workshops, pp. 2113–2119 (2011)
6. Dziwiński, P., Bartczuk, Ł., Przybył, A., Avedyan, E.D.: A new algorithm for identification of significant operating points using swarm intelligence. In: Rutkowski, L., Korytkowski, M., Scherer, R., Tadeusiewicz, R., Zadeh, L.A., Zurada, J.M. (eds.) ICAISC 2014. LNCS (LNAI), vol. 8468, pp. 349–362. Springer, Heidelberg (2014). doi:10.1007/978-3-319-07176-3_31
7. Huang, D., Ardabilian, M., Wang, Y., Chen, L.: Asymmetric 3D/2D face recognition based on LBP facial representation and canonical correlation analysis. In: International Conference on Image Processing, pp. 3325–3328 (2009)

8. Kanungo, T.: Hidden Markov Model Tutorial (1999). http://www.kanungo.com/software/hmmtut.pdf
9. Kompanets, L.: Biometrics of asymmetrical face. In: Zhang, D., Jain, A.K. (eds.) ICBA 2004. LNCS, vol. 3072, pp. 67–73. Springer, Heidelberg (2004). doi:10.1007/978-3-540-25948-0_10
10. Kubanek, M., Rydzek, S.: A hybrid method of user identification with use independent speech and facial asymmetry. In: Rutkowski, L., Tadeusiewicz, R., Zadeh, L.A., Zurada, J.M. (eds.) ICAISC 2008. LNCS (LNAI), vol. 5097, pp. 818–827. Springer, Heidelberg (2008). doi:10.1007/978-3-540-69731-2_78
11. Kurach, D., Rutkowska, D.: Influence of facial asymmetry on human recognition. In: Rutkowski, L., Korytkowski, M., Scherer, R., Tadeusiewicz, R., Zadeh, L.A., Zurada, J.M. (eds.) ICAISC 2012. LNCS (LNAI), vol. 7268, pp. 276–283. Springer, Heidelberg (2012). doi:10.1007/978-3-642-29350-4_33
12. Mahoor, M., Abdel-Mottaleb, M.: Face recognition based on 3D ridge images obtained from range data. Pattern Recogn. 42(3), 445–451 (2009)
13. Mitra, S., Lazar, N.A., Liu, Y.: Understanding the role of facial asymmetry in human face identification. J. Stat. Comput. 17(1), 57–70 (2007)
14. Ostwald, J., Berssenbrggea, P., Dirksena, D., Runtea, C., Wermkerb, K., Kleinheinzc, J., Jungc, S.: Measured symmetry of facial 3D shape and perceived facial symmetry and attractiveness before and after orthognathic surgery. J. Craniomaxillofac. Surg. 43(4), 521–527 (2015)
15. Rama, A., Tarres, F., Onofrio, D., Tubaro, S.: Mixed 2D–3D information for pose estimation and face recognition. ICASSP II, 361–368 (2006)
16. Riccio, D., Dugelay, J.-L.: Asymmetric 3D/2D processing: a novel approach for face recognition. In: Roli, F., Vitulano, S. (eds.) ICIAP 2005. LNCS, vol. 3617, pp. 986–993. Springer, Heidelberg (2005). doi:10.1007/11553595_121
17. Samaria, F., Young, S.: HMM-based architecture for face identification. Image Vis. Comput. 12(8), 537–583 (1994)
18. Yang, W., Yi, D., Lei, Z., Sang, J., Li, S.: 2D–3D face matching using CCA. In: 8th IEEE International Conference on Automatic Face & Gesture Recognition, FG 2008, pp. 1–6 (2008)
19. Zhang, G., Wang, Y.: Asymmetry-based quality assessment of face images. In: Bebis, G., et al. (eds.) ISVC 2009. LNCS, vol. 5876, pp. 499–508. Springer, Heidelberg (2009). doi:10.1007/978-3-642-10520-3_47
20. Zhao, W., Chellappa, R., Phillips, P., Rosenfeld, A.: Face recognition: a literature survey. ACM Comput. Surv. 35(4), 399–458 (2003)

# Best-Fit Segmentation Created Using Flood-Based Iterative Thinning

Adam Piórkowski$^{(\boxtimes)}$

Department of Geoinfomatics and Applied Computer Science,
AGH University of Science and Technology,
A. Mickiewicza 30 Av., 30–059 Krakow, Poland
pioro@agh.edu.pl

**Abstract.** Classical methods of segmentation that use binarization in the preprocessing stage often do not provide the precise delineation of the range of objects. For example, this might be useful for images of the corneal endothelium obtained with specular or confocal microscopy. This article presents a solution that makes it possible to adjust the course of the segmentation in the valleys between cells. The algorithm is a combination of iterative thinning and a watershed algorithm that works by the gradual removal of points with increasingly lower brightness levels. The article also contains examples of output images and quality tests.

**Keywords:** Segmentation · Iterative thinning · Corneal endothelium

## 1 Introduction

The segmentation of corneal endothelium images is a difficult issue due to the nature of the images obtained with confocal or specular microscopy. These images have a relatively high proportion of thermal noise, low contrast, and non-uniform luminance, all of which are qualities that prevent the use of algorithms such as the watershed algorithm. Instead, there are different methods to obtain binary images.

The authors in [12] propose the use of non-subsampled wavelet pyramid decomposition of lowpass regions. The pre-processing includes removal of non-uniform illumination and noise. Post processing includes dilatation, erosion, opening and closing operations, and finally thinning. Other method to deal with blown-out illumination areas is presented in [16]. Mahzoun et al. [13] propose using six convolution masks of size $9 \times 9$ that perform directional filtering (vertical, horizontal, left and right) and custom filtering (two 'tricorn' masks and binarization of the sum of outputs). In [9], a similar effective method is presented which assumes the use of four directional masks. These two approaches need thinning as the last step of segmentation. Sanches [21] uses scale-space filtering, especially Gaussian, to make use of the separability property of the Gaussian kernel, and unsharp masking. Adaptive thresholding and binarization are also used. Subsequently, skeletonization (thinning) is performed.

© Springer International Publishing AG 2017
R.S. Choraś (ed.), *Image Processing and Communications Challenges 8*,
Advances in Intelligent Systems and Computing 525, DOI 10.1007/978-3-319-47274-4_7

A custom method is presented in [15]. The authors propose a scissoring operator that separates cells in the binary image. The most advanced approaches use the active contour technique to obtain the shape of each cell in the image [1,10], or artificial neuron networks with numerical filters designed for the border extraction problem [7].

## 2    The Problem of Imprecise Segmentation

Most of the presented approaches (even clinical programs) do not necessarily perform precise segmentation. The true shape of some cells is not precisely drawn (Fig. 1). This problem has been described in the literature [2,11,14,18].

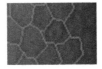

**Fig. 1.** A sample of imprecise segmentation

### 2.1    Thinning

In all these approaches, a binary image is thinned. Iterative thinning is usually the gradual removal of points from all sides of objects [20]. When images of the corneal endothelium are binarized, a side effect of this approach is that the dividing line between the cells does not necessarily fall on the pits or valleys. Some dividing lines are located on the slopes of particular cells. Therefore, this does not make for a representative and repetitive segmentation as some of cell grid parameters are sensitive and can depend on the segmentation method. To ensure the objectivity and reproducibility of clinical parameters, the segmentation lines should obtained, as they ran at the bottom of valleys.

The lack of precision may occur during the thinning of the binary images that are the result of most of the presented algorithms. In the classic approach to thinning, iterative symmetrical cuts are made to binary images by deleting points evenly on each side of the object to obtain a contour that is one pixel thick. This approach may cause the resulting contour not to cover the bottom of the valley (Fig. 2).

Thinning is usually carried out using a Hit-or-Miss algorithm with a set of carefully selected masks (Golay Alph.). In the present work, the set of masks described in Table 1 are used; each of them appears in orientations of $0°$, $90°$ $180°$ and $270°$.

To solve the problem of imprecise thinning (imprecise segmentation) some approaches have been developed.

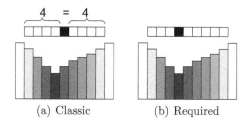

(a) Classic                    (b) Required

**Fig. 2.** Differences between classic and required thinning

**Table 1.** A set of masks for thinning

$$\begin{bmatrix} X & 1 & 1 \\ 0 & 1 & 1 \\ 0 & 0 & X \end{bmatrix} \begin{bmatrix} X & 1 & X \\ X & 1 & X \\ 0 & 0 & 0 \end{bmatrix} \begin{bmatrix} X & X & 0 \\ 0 & 1 & 0 \\ 0 & 0 & 0 \end{bmatrix} \begin{bmatrix} 0 & X & X \\ 0 & 1 & 0 \\ 0 & 0 & 0 \end{bmatrix}$$

## 2.2 Adjusting the Valley Segmentation Courses

Very interesting solution to the problem of search segmentation best suited to valleys was presented by B. Selig et al. [22]. The method comprises two parts. In the first stage, the seeded watershed is applied many times with seeds placed randomly (stochastic watershed). The next step blurs the output with a Gaussian filter, which removes local minima, and finally a classic watershed is used. To evaluate the output, the mean gray values of source pixels under the segmentation line (MGV) are calculated.

A second approach is presented in [18], the first part of which determines the number of neighbors of less than or equal brightness to each image point in the input image. The second phase performs iterative thinning on the basis of the constructed neighborhood map (NMIT) [17]. Each of the 9 thinning iterations removes only points that correspond to values in the maps.

## 3   Flood-Based Iterative Thinning Algorithm (FIT)

In order to solve the problem, a new algorithm is proposed which is a combination of a thinning algorithm and a watershed algorithm. This approach involves iterative thinning of binary images with reference to the source input image. Thinning iterations run in the same way as flooding in the watershed algorithm; in the case of dark valleys and bright objects in the source image, the iterations are carried out from the highest brightness (255 for 8 bit grayscale images) to the lowest brightness (0). In each iteration, only points of the binary image where the brightness in the source image is greater than or equal to the number of the iteration are processed. As a result, the highest points above and the lowest points below are removed, unless, of course, they match the thinning mask pattern. This method reduces the symmetrical thinning phenomenon; in its place, there is a solution most suited to the valleys in the image. A similar approach of thinning for grayscale images is presented in [3,4]. The algorithm can be presented in the form of pseudo-code:

```
FOR level = 255 TO 0
    ThinPoints(segmentation) WHERE SourceImage[x,y] >= level
```

### 3.1   Improvement of Thinning

The resulting segmentation of the binary image might be not necessarily the best fit because the binary image could cover a slope, instead of the lowest parts of the valley. In this case, the proposed solution is a cyclical correction, based on dilation and iterative thinning. No more improvements are made when there are no more changes, or a looped cycle is detected.

### 3.2   Testing of Flood-Based Iterative Thinning

Tests were carried out to assess a sample of segmentation quality. As a source data, the Endothelial Cell Alizarine Data Set was chosen [19]. It contains 30 corneal endothelium images and the corresponding manual segmentations.

Figure 3(a) contains a fragment of the original image and manual seed (Fig. 3(b)) from the data set (No 4). The next figure (Fig. 3(c)) shows segmentation seed after dilatation. The drawings (Fig. 4) show a selection of iterative thinning stages for levels 140, 120, 100, 90, 80 and 60. It can be seen that this iterative thinning first removes points that cover the higher parts, and then deletes the lowest points.

A qualitative assessment of the proposed method is needed. For this purpose, a comparison of the original thinned segmentation and NMIT and FIT methods was prepared. Segmentation was corrected by performing iterative cycles of dilatation and modified thinning (Fig. 5). A description of the subsequent stage, including the differences between images and the mean gray value of segmentation (MGV), are presented in Table 2, which shows the results for the thinned original segmentation, improved by NMIT and FIT algorithms.

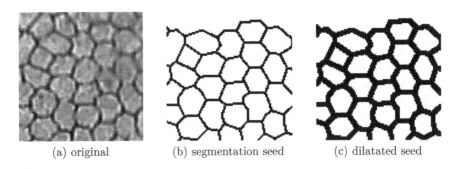

(a) original          (b) segmentation seed          (c) dilatated seed

**Fig. 3.** Input pictures

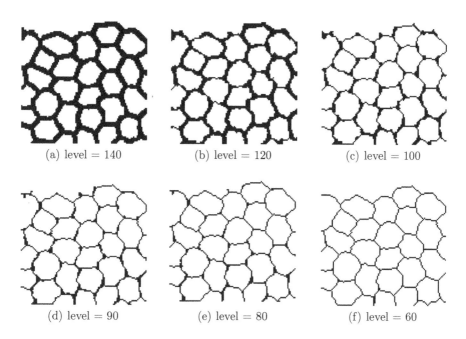

(a) level = 140          (b) level = 120          (c) level = 100

(d) level = 90           (e) level = 80           (f) level = 60

**Fig. 4.** Thinning iterations for selected levels

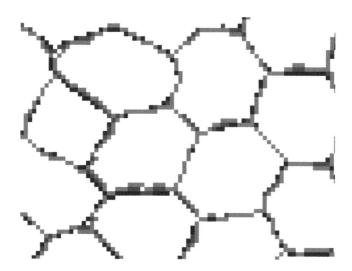

**Fig. 5.** Differences between segmentations obtained in subsequent dilatation and FIT thinning cycles; black - initial segmentation, blue (dark gray) - first cycle, red (light gray) - second cycle (final) (Color figure online)

**Table 2.** The performance of iterative thinnings

| Cycle | NMIT | | FIT | |
|---|---|---|---|---|
| | differences [px] | MGV | differences [px] | MGV |
| Initial thinning | | 112.7477 | | 112.7477 |
| 1 | 9587 | 109.4422 | 10152 | 109.2717 |
| 2 | 807 | 109.3556 | 467 | 109.2190 |
| 3 | 94 | 109.3536 | 0 | 109.2190 |
| 4 | 24 | 109.3413 | | |
| 5 | 8 | 109.3363 | | |
| 6 | 0 | 109.3363 | | |

## 4    Conclusions

The presented algorithm matched the lines of segmentation to the source image fairly well. In addition, not only the optical effect Fig. 6 can be recognized as the obtained mean gray value of segmentation was better for the FIT algorithm than the NMIT algorithm and classical thinning. Moreover, the number of cycles of dilatation and thinning was smaller. It can be concluded that the proposed algorithm performs solves the problem quite well. Further studies will focus on the quality of thinning compared to other methods (eg. FRSW) and the effects of the method for changing shape factors, eg. $CV$ [5], $H$ [6] and $CVSL$ [8].

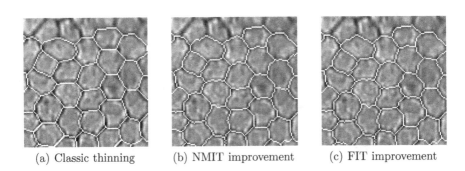

(a) Classic thinning        (b) NMIT improvement        (c) FIT improvement

**Fig. 6.** Input pictures with overlayed segmentation

**Acknowledgement.** This work was financed by the AGH - University of Science and Technology, Faculty of Geology, Geophysics and Environmental Protection as a part of statutory project.

# References

1. Charłampowicz, K., Reska, D., Boldak, C.: Automatic segmentation of corneal endothelial cells using active contours. Adv. Comput. Sci. Res. **11**, 47–60 (2014)
2. Cheung, S.W., Cho, P.: Endothelial cell analysis using the Topcon SP-1000 non-contact specular microscope and IMAGEnet system. Clin. Exp. Optom. **81**(1), 1–7 (1998)
3. Couprie, M., Bezerra, F.N., Bertrand, G.: Topological operators for grayscale image processing. J. Electron. Imaging **10**(4), 1003–1015 (2001)
4. Couprie, M., Bezerra, N., Bertrand, G.: A parallel thinning algorithm for grayscale images. In: Gonzalez-Diaz, R., Jimenez, M.-J., Medrano, B. (eds.) DGCI 2013. LNCS, vol. 7749, pp. 71–82. Springer, Heidelberg (2013). doi:10.1007/978-3-642-37067-0_7
5. Doughty, M.: The ambiguous coefficient of variation: polymegethism of the corneal endothelium and central corneal thickness. Int. Cont. Lens Clin. **17**(9–10), 240–248 (1990)
6. Doughty, M.: Concerning the symmetry of the 'hexagonal' cells of the corneal endothelium. Exp. Eye Res. **55**(1), 145–154 (1992)
7. Foracchia, M., Ruggeri, A.: Cell contour detection in corneal endothelium in-vivo microscopy. In: Proceedings of the 22nd Annual International Conference of the IEEE on Engineering in Medicine and Biology Society, vol. 2, pp. 1033–1035. IEEE (2000)
8. Gronkowska-Serafin, J., Piorkowski, A.: Corneal endothelial grid structure factor based on coefficient of variation of the cell sides lengths. In: Choras, R.S. (ed.) Image Processing and Communications Challenges 5. Advances in Intelligent Systems and Computing, vol. 233, pp. 13–19. Springer, Heidelberg (2014)
9. Habrat, K., Habrat, M., Gronkowska-Serafin, J., Piórkowski, A.: Cell detection in corneal endothelial images using directional filters. In: Choraś, R.S. (ed.) Image Processing and Communications Challenges 7. AISC, vol. 389, pp. 113–123. Springer, Heidelberg (2016). doi:10.1007/978-3-319-23814-2_14
10. Hachaj, T., Ogiela, M.R.: Application of centerline detection and deformable contours algorithms to segmenting the carotid lumen. J. Electron. Imaging **23**(2), 023006–023006 (2014)
11. Jaworek-Korjakowska, J., Tadeusiewicz, R.: Design of a teledermatology system to support the consultation of dermoscopic cases using mobile technologies and cloud platform. Bio-Algorithms Med-Syst. **11**(1), 53–58 (2015)
12. Khan, M.A.U., Niazi, M.K.K., Khan, M.A., Ibrahim, M.T.: Endothelial cell image enhancement using non-subsampled image pyramid. Inf. Technol. J. **6**(7), 1057–1062 (2007)
13. Mahzoun, M., Okazaki, K., Mitsumoto, H., Kawai, H., Sato, Y., Tamura, S., Kani, K.: Detection and complement of hexagonal borders in corneal endothelial cell image. Med. Imaging Technol. **14**(1), 56–69 (1996)
14. Mazurek, P., Oszutowska-Mazurek, D.: From the slit-island method to the ising model: analysis of irregular grayscale objects. Int. J. Appl. Math. Comput. Sci. **24**(1), 49–63 (2014)
15. Nadachi, R., Nunokawa, K.: Automated corneal endothelial cell analysis. In: Proceedings of Fifth Annual IEEE Symposium on Computer-Based Medical Systems, pp. 450–457. IEEE (1992)

16. Nurzyńska, K., Haraszczuk, R.: Detection and normalization of blown-out illumination areas in grey-scale images. In: Bebis, G., Boyle, R., Parvin, B., Koracin, D., Fowlkes, C., Wang, S., Choi, M.-H., Mantler, S., Schulze, J., Acevedo, D., Mueller, K., Papka, M. (eds.) ISVC 2012. LNCS, vol. 7431, pp. 282–291. Springer, Heidelberg (2012). doi:10.1007/978-3-642-33179-4_28

17. Piórkowski, A.: A statistical dominance algorithm for edge detection and segmentation of medical images. In: Pietka, E., Badura, P., Kawa, J., Wieclawek, W. (eds.) Information Technologies in Medicine. AISC, vol. 471, pp. 3–14. Springer, Heidelberg (2016). doi:10.1007/978-3-319-39796-2_1

18. Piórkowski, A., Gronkowska–Serafin, J.: Towards precise segmentation of corneal endothelial cells. In: Ortuño, F., Rojas, I. (eds.) IWBBIO 2015. LNCS, vol. 9043, pp. 240–249. Springer, Heidelberg (2015). doi:10.1007/978-3-319-16483-0_25

19. Ruggeri, A., Scarpa, F., De Luca, M., Meltendorf, C., Schroeter, J.: A system for the automatic estimation of morphometric parameters of corneal endothelium in alizarine red-stained images. Br. J. Ophthalmol. 94(5), 643–647 (2010)

20. Saeed, K., Tabędzki, M., Rybnik, M., Adamski, M.: K3M: a universal algorithm for image skeletonization and a review of thinning techniques. Int. J. Appl. Math. Comput. Sci. 20(2), 317–335 (2010)

21. Sanchez-Marin, F.: Automatic segmentation of contours of corneal cells. Comput. Biol. Med. 29(4), 243–258 (1999)

22. Selig, B., Vermeer, K.A., Rieger, B., Hillenaar, T., Hendriks, C.L.L.: Fully automatic evaluation of the corneal endothelium from in vivo confocal microscopy. BMC Med. Imaging 15(1), 1 (2015)

# A Comparative Study of Image Enhancement Methods in Tree-Ring Analysis

Anna Fabijańska[1]([⊠]), Małgorzata Danek[2], Joanna Barniak[3],
and Adam Piórkowski[4]

[1] Institute of Applied Computer Science, Lodz University of Technology,
18/22 Stefanowskiego Street, 90–924 Lodz, Poland
anna.fabijanska@p.lodz.pl

[2] Department of Environmental Analysis, Mapping and Economic Geology,
AGH University of Science and Technology,
A. Mickiewicza 30 Av., 30–059 Krakow, Poland
mdanek@agh.edu.pl

[3] Department of General Geology and Geotourism,
AGH University of Science and Technology,
A. Mickiewicza 30 Av., 30–059 Krakow, Poland
barniak@geol.agh.edu.pl

[4] Department of Geoinfomatics and Applied Computer Science,
AGH University of Science and Technology,
A. Mickiewicza 30 Av., 30–059 Krakow, Poland
pioro@agh.edu.pl

**Abstract.** In this paper the problem of semiautomatic tree-ring detection in scanned images of European larch wood sample is considered. In particular, the attention is paid to find image enhancement approach which increases the number of tree rings detected in the wood image by the CooRecorder software. The results provided by different preprocessing methods (including thresholding, contrast enhancement and various spatial filters) are assessed by means of the number of the detected tree rings. Discussion and interpretation of the results are also provided.

**Keywords:** Edge detection · Segmentation · Image processing · Preprocessing · Tree-ring analysis · European larch

## 1 Introduction

Dendrochronology is a science that analyses information contained in so called tree rings, i.e. annual wood increment. This information can next be used in many other fields of science (e.g. archeology, geology, climatology, ecology, geomorphology or glaciology). The main feature of tree rings that is used in the above mentioned analysis is their width. The classical tree-ring width measurement bases on prepared pieces or cores of wood. Tree rings are measured one by one by the operator using special equipment: a microscope, a sliding table and a data recorder. The alternative way of measurement is based on

© Springer International Publishing AG 2017
R.S. Choraś (ed.), *Image Processing and Communications Challenges 8*,
Advances in Intelligent Systems and Computing 525, DOI 10.1007/978-3-319-47274-4_8

scans or photographs of wood samples. In such a case, tree rings are measured in digital images using dedicated programs (e.g. CDendro&CooRecorder, WinDENDRO™, LignoVision™). Nevertheless, using the microscope still seems to be the most accurate way of measurement. However, it is time consuming and requires expensive equipment.

One of the advantages of using wood scans is the automation of measurements, that can reduce the measurement time radically. The possibility and accuracy of the image based measurements depend on the quality of the scans and appropriate sample preparation method. It is also influenced by the anatomical features of the wood. Since different species have differently complicated wood structure, it can influence the visibility of the tree-ring boundaries. As a result, automatic detection of tree-ring boundaries on scans can be very difficult even if these are easily distinguishable for an expert. That is why recently image processing based methods have been undergoing constant improvement (e.g. [5,6,10]). Also a study presented in this paper is a step forward in this direction. In particular, this study focuses on finding the most proper image preprocessing methods for semiautomatic tree-ring measurements of European larch (*Larix decidua* Mill). In this case, the preprocessing is aimed at highlighting image information at tree rings and thus enable (or improve) semiautomatic dendrochronogical measurements performed in the dedicated CooRecorder software.

European larch represents relatively not complicated anatomical structure of coniferous wood type, in which each tree ring contains of two layers: the lighter one early wood and the darker one (late wood). Even in the case of such a structure, the semiautomatic measurements done on raw scans can be unreliable. The problems occur especially if the tree rings are very narrow, the darker part of the ring (late wood) is rather pale or the set line of measurement cross the transition zone between sapwood and heartwood (see Fig. 1(a)). Therefore, in this paper different approaches to image preprocessing are tested in order to find the one which increases the number of the rings detected semiautomatically by CooRecorder.

## 2  Attempts to Enhancement of Wood Core Images

### 2.1  Thresholding

The first group of methods applied for tree rings enhancement was thresholding. Both the global and the local threshold selection were tested. The global threshold was selected manually in order to obtain the best result. Local thresholding was performed in the window of radius $R = 25$ pixels with regard to mean and isodata threshold [2] determined for each window. The thresholding was performed separately for each channel in the HSV color space. Then, the best result was selected. The results of thresholding performed as described above are presented in Fig. 2. The tree rings detected in the corresponding images are indicated by numbered cross markers.

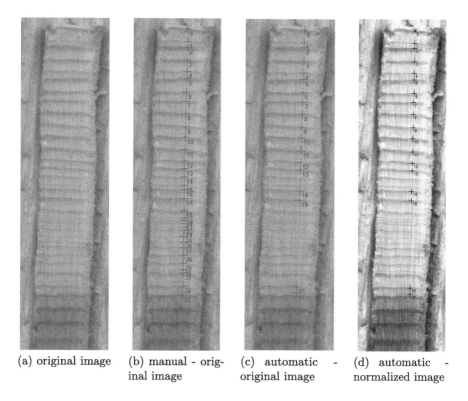

(a) original image      (b) manual - orig-      (c) automatic    -      (d) automatic    -
                        inal image             original image          normalized image

**Fig. 1.** The results of tree-ring detection

## 2.2   Contrast Enhancement Methods

The second attempt to highlighting tree rings was contrast manipulation. Firstly, contrast stretching (see Fig. 1(d)) and contrast equalization (see Fig. 3(a)) were applied to the global histogram of the image. This however did not provide rewarding results. Therefore, in the next step Contrast Limited Adaptive Histogram Equalisation (CLAHE) was applied [8,12]. The method transforms pixel intensity with respect to its neighbourhood. In order to avoid overamplification of noise the local contrast is limited by the slope of the transformation function. The size of the neighbourhood and the slope of the transformation function are two parameters of the CLAHE approach. The results of Contrast Limited Adaptive Histogram Equalisation applied to a sample image of larch with an increasing slope are shown in Fig. 3(b), (c), (d). These were obtained for neighbourhood size equal to 127 pixels. Again, the tree rings detected in the corresponding images are indicated by numbered cross markers.

## 2.3   Textural Features

The analysis of textural features was another approach considered in the pre-processing step. In particular, Haralick features [1,4] were determined for input

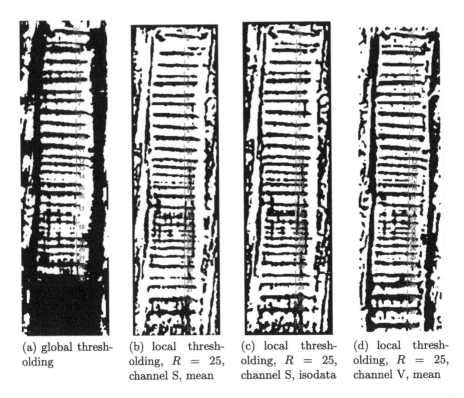

(a) global thresh-olding

(b) local thresh-olding, $R = 25$, channel S, mean

(c) local thresh-olding, $R = 25$, channel S, isodata

(d) local thresh-olding, $R = 25$, channel V, mean

**Fig. 2.** The results of tree-ring detection in a larch strip image thresholded with different methods

image. Filtration was preformed in each channel of the HSV color space separately and then combined into a colour image. The window of size $5 \times 5$ pixels was used. The results of filtration performed with respect to different features are shown in Fig. 4. Only the features which provided reasonable results were selected for presentation.

### 2.4 Using Convolution Filters

Another considered approach to detect the tree-ring boundaries was the use of different directional convolution filters [9,11]. The most interesting result was obtained for the filter described by mask 1. The resulting image is shown in Fig. 5(a).

$$\begin{bmatrix} 3 & 3 & 3 & 3 & 3 & 3 & 3 \\ 2 & 2 & 2 & 2 & 2 & 2 & 2 \\ 1 & 1 & 1 & 1 & 1 & 1 & 1 \\ 0 & 0 & 0 & 0 & 0 & 0 & 0 \\ -1 & -1 & -1 & -1 & -1 & -1 & -1 \\ -2 & -2 & -2 & -2 & -2 & -2 & -2 \\ -3 & -3 & -3 & -3 & -3 & -3 & -3 \end{bmatrix} \tag{1}$$

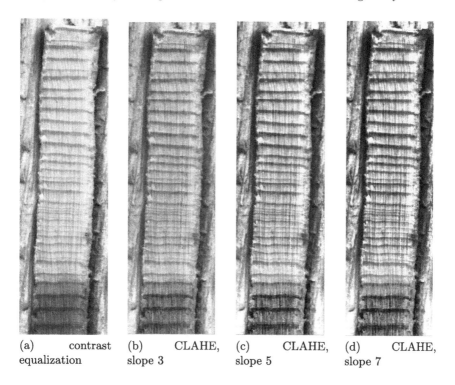

(a)    contrast    (b)    CLAHE,    (c)    CLAHE,    (d)    CLAHE,
equalization    slope 3    slope 5    slope 7

**Fig. 3.** The results of tree-ring detection in contrast enhanced larch strip image

Another approach worth mentioning is a filter that determines the statistical dominance of pixels SDA [7]. In this case lighter pixels (early wood) dominate over the darker ones (late wood). This filter is designed primarily for the neighborhood with a round shape but for the purposes of wood cores processing is reduced to a directional filter. It allows to concentrate on increments, reducing the influence of other disturbances. The algorithm can be defined by the formula 2 (directional version):

$$p'(x,y) = \sum_{j=-N}^{j=N} \begin{cases} p'(x,y) := p'(x,y) + 1, & p(x,y+j) \geq p(x,y) + t, \\ & j^2 \leq R^2 \\ p'(x,y) := p'(x,y), & \text{otherwise} \end{cases} \tag{2}$$

where:

- $p(x,y)$ - the value of pixel $(x,y)$ in input image,
- $p'(x,y)$ - the value of pixel $(x,y)$ in output image,
- $R$ - radius of a neighborhood,
- $N$ - size of neighborhood mask, $N = \lceil R \rceil$
- $t$ - *threshold* - the optional difference to be checked.

The algorithm may be preceded by a preprocessing which aims at background compensation, like solution presented in [3]. It increases intensity of each pixel

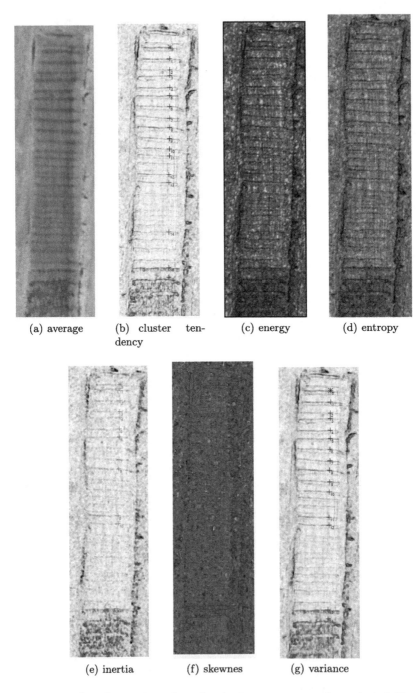

(a) average    (b) cluster ten-    (c) energy    (d) entropy
dency

(e) inertia    (f) skewnes    (g) variance

**Fig. 4.** The results of tree-rings detection in images representing selected Haralick textural features of a sample image of a larch strip

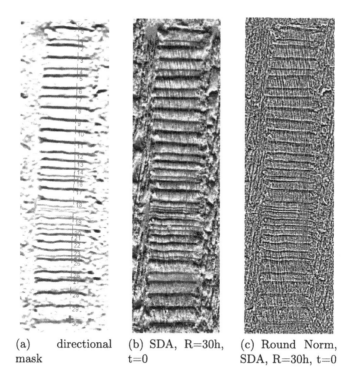

(a)     directional     (b) SDA, R=30h,     (c) Round Norm,
mask                    t=0                 SDA, R=30h, t=0

**Fig. 5.** The results of tree-rings detection in images a larch core processed with selected convolution filters

by a difference between the average image intensity and the average of intensity within a round, $r$-pixels sized neighborhood of this pixel. This is described by formula 3.

$$Pout_{ij} = Pin_{ij} + (S - Sr_{ij}) \qquad (3)$$

where:

- $Pin$ - input image,
- $Pout$ - output image,
- $S$ - arithmetic mean of intensity of all image pixels,
- $Sr_{ij}$ - arithmetic mean intensity for all pixels in the distance equal or less than $r$ from pixel $i, j$.

## 3   The Assessment of the Results

The Table 1 summarises the performance of image preprocessing techniques considered in this paper. In particular, for each method (which name is indicated in the column caption) the number of the detected tree rings is given and compared both: with the results of manual detection performed by a dendrochronologist (see Fig. 1(b)) and the semiautomatic detection obtained in CooRecorder software (see Fig. 1(c)). Additionally, for each case the number of correctly detected rings, incorrect rings and missing rings is given.

**Table 1.** The assessment of tree-ring detection results obtained for different preprocessing methods

| Method | Manual | Automatic | Normalized | Global thresholding | Local thresholding, S, m | Local thresholding, S, i | Local thresholding, V, m | Contrast equalization | CLAHE, slope 3 | CLAHE, slope 5 | CLAHE, slope 7 | Haralick, average | Haralick, cluster tendency | Haralick, energy | Haralick, entropy | Haralick, intertia | Haralick, skewnes | Haralick, variance | Directional | SDA | Round Norm, SDA |
|---|---|---|---|---|---|---|---|---|---|---|---|---|---|---|---|---|---|---|---|---|---|
| Detected | 37 | 17 | 17 | 23 | 29 | 31 | 31 | 17 | 21 | 24 | 25 | 4 | 16 | 33 | 30 | 13 | 28 | 14 | 30 | 32 | 37 |
| Correct | 37 | 16 | 16 | 23 | 29 | 30 | 31 | 16 | 21 | 23 | 24 | 3 | 15 | 27 | 25 | 11 | 19 | 13 | 30 | 32 | 34 |
| Poor accuracy | 0 | 0 | 2 | 3 | 2 | 1 | 0 | 0 | 0 | 0 | 0 | 3 | 0 | 1 | 1 | 0 | 8 | 0 | 1 | 0 | 0 |
| Incorrect | 0 | 1 | 1 | 0 | 0 | 1 | 0 | 1 | 0 | 1 | 1 | 1 | 1 | 7 | 5 | 2 | 9 | 1 | 0 | 0 | 3 |
| Missing | 0 | 21 | 21 | 14 | 8 | 7 | 6 | 21 | 16 | 14 | 13 | 34 | 22 | 10 | 12 | 26 | 18 | 24 | 7 | 5 | 3 |

The results presented in Table 1 and Fig. 1 show that the number of automatically detected tree-ring boundaries in original (unprocessed) image is low (16 out of 37). Application of global normalization and contrast equalization does not improve automatic detection success rate.

Among thresholding methods the best results were obtained for the local thresholding of V channel performed with regard to average intensity value (see Fig. 2). In this case, the number of correctly detected boundaries increased to 31. Among other thresholding methods poor accuracy and incorrectness in detection were observed, hence in the case of threshold-based methods, the precision of the automatic measurements requires further study.

In the case of contrast enhancement approaches (see Fig. 3) the number of correctly detected boundaries was lower, but there was no problem with precision and accuracy. It seems that in unprocessed, normalized, equalized and contrast enhanced images the sapwood/heart wood transition zone causes problems during automatic detection due to the large area color change.

Unsatisfying results were obtained also for Haralick textural features. For average the tree-ring boundaries were blurred (see Fig. 4(a)), similarly to directional mask (see Fig. 1). In the case of other considered features, CooRecorder detected many nonexistent (incorrect) tree-ring boundaries, especially for energy, entropy and skewness. There was also the problem with poor accuracy of detection (especially for skewness).

Among presented methods of preprocessing the best results were obtained for one of the convolution filters methods, namely SDA ($R = 30$ h, $t = 0$; Fig. 5(b)), but even in this case the full success rate was not achieved (32 out of 37 tree-ring boundaries were detected correctly, with no incorrect rings).

## 4    Conclusions

The study presented in this paper clearly shows, that preporocessing of wood core images increases the number of the tree rings detected by CooRecorder software. In the case of the best result obtained in this study only 5 tree rings out of 37 (c.a. 13 %) were missed and there were no incorrect tree rings at the same time. This is a significant improvement when compared to the measurement performed on the original image of wood strip where the tree-ring detection rate was only about 43 %. However, on the other hand it can be seen that none of the considered image preprocessing methods works perfectly. Even in the best case the results still can be improved, since some tree rings are missing or falsely added. Therefore there still is a need to develop dedicated image preprocessing methods which will emphasize existing tree rings and avoid detection of non-existing ones. Development of such a method will be the main goal of our future works and the first step to the development of software for fast processing of wood sample images dedicated for dendrochronological measurements.

**Acknowledgement.** This work was co-financed by the Lodz University of Technology, Faculty of Electrical, Electronic, Computer and Control Engineering as a part of statutory project no 501/12-24-2-5416.

This work was co-financed by the AGH - University of Science and Technology, Faculty of Geology, Geophysics and Environmental Protection as a part of statutory projects no. 11.11.140.173, no. 11.11.140.613 and no. 11.11.140.626.

## References

1. Chandy, D.A., Johnson, J.S., Selvan, S.E.: Texture feature extraction using gray level statistical matrix for content-based mammogram retrieval. Multimedia Tools Appl. **72**(2), 2011–2024 (2014)
2. El-Zaart, A.: Images thresholding using isodata technique with gamma distribution. Pattern Recogn. Image Anal. **20**(1), 29–41 (2010)
3. Habrat, K., Habrat, M., Gronkowska-Serafin, J., Piórkowski, A.: Cell detection in corneal endothelial images using directional filters. In: Choraś, R.S. (ed.) Image Processing and Communications Challenges 7. AISC, vol. 389, pp. 113–123. Springer, Heidelberg (2016). doi:10.1007/978-3-319-23814-2_14
4. Haralick, R.M., Shanmugam, K., Dinstein, I.: Textural features for image classification. IEEE Trans. Syst. Man Cybern. **SMC−3**(6), 610–621 (1973)
5. Laggoune, H., Guesdon, V., et al.: Tree ring analysis. In: Canadian Conference on Electrical and Computer Engineering, pp. 1574–1577. IEEE (2005)
6. Norell, K.: Automatic counting of annual rings on pinus sylvestris end faces in sawmill industry. Comput. Electron. Agric. **75**(2), 231–237 (2011)
7. Piórkowski, A.: A statistical dominance algorithm for edge detection and segmentation of medical images. In: Piętka, E., Badura, P., Kawa, J., Wieclawek, W. (eds.) Information Technologies in Medicine. AISC, vol. 471, pp. 3–14. Springer, Heidelberg (2016). doi:10.1007/978-3-319-39796-2_1
8. Reza, A.M.: Realization of the contrast limited adaptive histogram equalization (clahe) for real-time image enhancement. J. VLSI Sig. Process. Syst. **38**(1), 35–44 (2004)

9. Serra, J., Vincent, L.: An overview of morphological filtering. Circ. Syst. Sig. Process. **11**(1), 47–108 (1992)
10. Sundari, P.M., Kumar, S.B.R.: A study of image processing in analyzing tree ring structure. Int. J. Res. Humanit. Arts Lit. **2**(3), 13–18 (2014)
11. Tadeusiewicz, R., Korohoda, P.: Computer Analysis and Image Processing. Progress of Telecommunication Foundation Publishing House, Krakow (1997)
12. Zuiderveld, K.: Contrast limited adaptive histogram equalization. In: Heckbert, P.S. (ed.) Graphics Gems IV. Academic Press, Boston (1994)

# Key Frames Detection in Motion Capture Recordings Using Machine Learning Approaches

Tomasz Hachaj$^{(\boxtimes)}$

Institute of Computer Science, Pedagogical University of Krakow,
2 Podchorazych Ave, 30-084 Krakow, Poland
`tomekhachaj@o2.pl`

**Abstract.** The right choice of key frames is crucial for generation the valuable actions description for classification purposes. The machine learning approaches such as clustering of motion capture (MoCap) dataset can be used to calculate those key frames. K-means clustering already proved to be an efficient and effective method for key frames detection however to our knowledge no other clustering method has been used to this task. In this paper the several clustering methods namely model - based clustering with Gaussian Mixture model (Expectation - maximization algorithm), fuzzy clustering, K-medians clustering and hierarchical clustering are compared with K-means clustering in task of key frames detection. The comparison was done on dataset consisted of MoCap multimedia-quality recording of karate techniques that consisted of 12 different actions types (totally 480 actions samples). Results showed that the difference between clustering results increases while dataset is partitioned into more clusters. The second experiment was an evaluation what is an averaged percentage of recordings in training set and test set that contains all key frames detected by K-means algorithm. The proposed approach seems to have good generalization on the multimedia dataset when clusters number was set to two or three. The description of those two experiments is main novelty of this paper.

**Keywords:** Clustering · Key frames · Motion capture · Action recognition · Oyama Karate

## 1 Introduction

There are two most popular approaches to human action classification: the first is based on body joints trajectory analysis (for example Dynamic Time Warping) and the second on key frames detection. The detailed state of the art report on action recognition can be found in papers [7–9]. Key frames of a movement are important (from statistical point of view) part of an action that characterize all recordings of considered type. Also an action can be correctly classified to a given class if all key frames are present in its recording.

The right choice of key frames is crucial to generate the valuable actions description for classification purposes [3]. The machine learning approaches such

© Springer International Publishing AG 2017
R.S. Choraś (ed.), *Image Processing and Communications Challenges 8*,
Advances in Intelligent Systems and Computing 525, DOI 10.1007/978-3-319-47274-4_9

as clustering of motion capture (MoCap) dataset can be used to calculate the key frames. K-means clustering already proved to be an efficient and effective method for key frames detection [8] however to our knowledge no other clustering method has been used to this task. In this paper the several clustering methods are compared with K-means clustering in the task of key frames detection. The second experiment was an evaluation what is an averaged percentage of recordings in training set and test set that contains all key frames detected by K-means algorithm. The goal of this research was to check if key frames are well taught by the proposed approach and to check the generalization property of the proposed method - how well can it deal with a new data. The description of those two experiments is main novelty of this paper.

## 2   Materials and Methods

In this section we will present the features set we used in the experiment, the dataset, clustering algorithms that were evaluated and clustering results comparison methods.

### 2.1   Features Set

To gather the dataset for the experiment the multimedia MoCap system was used namely Kinect 2 and Microsoft Kinect SDK 2.0. It uses markerless body joints tracking technology with which 25 three-dimensional coordinates of body joints are detected and tracked with 30 Hz frequency. The initially detected features are dependent to body proportions of tracked user and to relative position of a person to camera. To make the actions independent to those factors we use feature set proposed in [10]. Movements are modeled in features space calculated from angles between vectors derived from tracked body joints. Those chosen vectors are visualized in Fig. 1. Angles are calculated between three vectors: $x$ (derived as difference between right and left shoulder coordinates), $y$ (derived as difference between coordinates of body joint between shoulders and between hips) and $z$ (a cross product of $x$ and $y$) and eight vectors $(v_1, .., v_8)$ which are shown as gray dotted lines.

The dataset we used is consisted of MoCap recordings of two professional sport (black belt) instructors and masters of Oyama Karate. Each experiment participant performed 12 different types of karate movements. Each of those movements was repeated 20 times. The whole dataset contained 260 movement samples per person (totally 480 samples). There were three types of defense techniques (Gedan-barai, Jodan-uke and Soto-uke - right hand only) three types of kicks (Hiza-geri, Mae-geri, Yoko-geri - right leg only) and three types of punches (Furi-uchi, Shita-uchi, Tsuki) done both left and right hand. The dataset was recorded on standard PC with application GDL Studio [1]. The features set that was used to model blocking techniques and right-hand punches are angles between vector $v1$ and $x$, $y$ and $z$ (three angles) and between $v_2$ and $x$, $y$ and

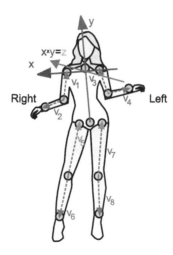

**Fig. 1.** This figure presents body joints set that was used to calculate vectors and to derive features set

$z$ (next three angles). To model left-hand punches and right-leg kicks I also calculated the angles between $x$, $y$ and $z$ but this time I used vector $v_3$ and $v_4$ for punches and $v_5$ and $v_6$ for kicks.

## 2.2 The Action's Key Frames Detection with Clustering Algorithms

The key frame definition can be derived from a cluster definition. For example a given MoCap frame belongs to a cluster (is similar to a key frame) if:

$$\left(\frac{x_0 - m_0}{sd_0 + \varepsilon_0}\right)^2 + ... + \left(\frac{x_n - m_n}{sd_n + \varepsilon_n}\right)^2 \leq 1 \tag{1}$$

Where:

$x_0, .., xn$ - feature values in a given MoCap frame;

$m_0, .., m_n$ - cluster center (mean values of coordinates of all elements assigned to a cluster)

$sd0, .., sd_n$ - cluster size (standard deviations of coordinates of all elements assigned to a cluster);

$\varepsilon_0, .., \varepsilon_n$ - epsilons values that extend clusters sizes (in case if variables distribution is not normal).

In Fig. 2 we present key frames of Gedan-barai block detected with K-means algorithm. The clustering has been done on all Gedan-barai MoCap recordings that came from the single person from dataset described in Sect. 2.1.

In previous researches [8,9] the K-means clustering was successfully used for key frames detection however other clustering algorithms has not been yet tested in this task. Due to this fact we decided to compare clustering obtained with other clustering algorithms with K-means results. We have chosen following popular algorithms for comparison:

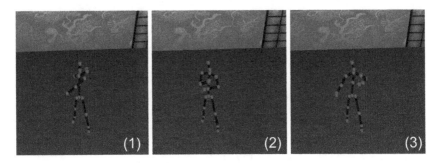

**Fig. 2.** This figure presents three key frames of Gedan-barai block. Those key frames were detected by proposed approach (1) with K-means clustering

- Model - based clustering with Gaussian Mixture model (Expectation - maximization algorithm) [5];
- Fuzzy clustering [11];
- K-medians clustering [11];
- Hierarchical (agglomerative) clustering (distances between clusters are recomputed by the Lance - Williams dissimilarity update formula, Ward's minimum variance method aims at finding spherical clusters [12]).

The density - based clustering is not applicable because there are no clusters that are defined as areas of higher density than the remainder of the data set. Also objects that are present in sparse regions should not be considered as clusters border points or outliers [4].

To generate the key frames set the clustering has to be done on large set of actions examples that should came from many persons for better averaging and generalization. The key frames detection is valid if detected key frames exists in all or majority of individual MoCap actions exemplars.

### 2.3 Clustering Results Evaluation

There are several popular methods that are used for clustering results comparison. The notation that will be use for methods definition is [13]:

$n_{11}$ - is a number of objects pairs that are in the same cluster under clustering C and clustering C';

$n_{00}$ - is a number of objects pairs that are in different clusters under clustering C and clustering C';

$n_{10}$ - is a number of objects pairs that are in the same cluster under clustering C but in different cluster under clustering C';

$n_{01}$ - is a number of objects pairs that are in different clusters under clustering C but in the same cluster under clustering C';

$n$ - is the total number of objects in clustered dataset.

The first comparison method we will use is a Rand index:

$$Rand(C, C') = \frac{2(n_{11} + n_{00})}{n(n-1)} \qquad (2)$$

It divides number of pairs that are the assigned in the same way in clustering C and C' by total number of all pairs. Rand index ranges from 0 where there is no pair classified in the same with in C and C' to 1 where clustering results are identical.

The Jaccard index [6]:

$$Jaccard(C, C') = \frac{n_{11}}{n_{11} + n_{10} + n_{01}} \qquad (3)$$

is computed by division of elements pairs that are in same clusters in C and C' by a sum of all elements pairs that are in same cluster in C and C' and objects pairs that are in different cluster in C and C'.

The Fowlkes - Mallows index:

$$Fowlkes - Mallows(C, C') = \frac{n_{11}}{\sqrt{(n_{11} + n_{10})(n_{11} + n_{01})}} \qquad (4)$$

It can be understood as geometric mean of precision and recall (geometric mean of exactness of C and C' and completeness of considered pairs in C and C').

## 3    Results

In this section comparison of clustering results of algorithms mentioned in Sect. 2.2 to K-means clustering results is presented. The evaluation was done on dataset described in Sect. 2.1. The clustering was performed on MoCap data from each action type separately. The data was partitioned into 2 to 5 clusters. Each experiment for each action type was repeated 100 times for each clustering method. The results were compared to K-means clustering. The averaged results plus/minus standard deviation are presented in Fig. 3. Due to paper space limitation we will present figure only.

Because clusters obtained with four tested algorithms and K-means did not differ much (it will be discussed in following section) the rest of the evaluation will be done only for k-means clustering. The second experiment was an evaluation what is an average percentage of recordings in training set and test set (Fig. 4) that contains all key frames detected by k-means algorithm. Each action type was evaluated separately. Training set contains MoCap data of first karate master while validation set MoCap data of the other person. Due to paper space limitation we will present figure only.

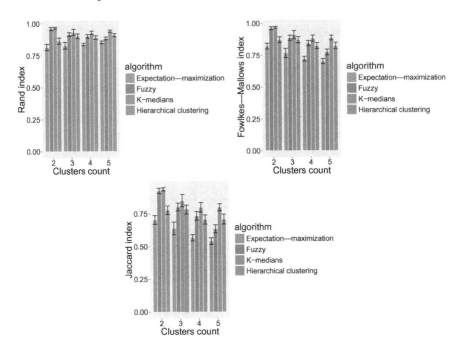

**Fig. 3.** This figure shows Rand, Fowlkes - Mallows and Jaccard index values for various clustering algorithms calculated from comparison with K-means clustering

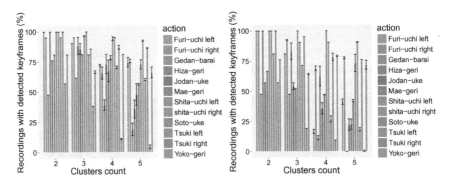

**Fig. 4.** This figure presents average percentage of recordings in the training set (left) and validation set (right) that contains all key frames detected by K-means algorithm. Black boxes visualize the standard deviation

## 4   Discussion and Conclusions

In the first experiment model - based clustering with Gaussian Mixture (expectation - maximization algorithm), fuzzy clustering, k-medians and hierarchical clustering results have been compared with k-means algorithm. As can be seen in Fig. 3 the Rand index does not differ much statistically while number of clusters changes. That means that sums of numbers of objects pairs that are in the same cluster under clustering C and clustering C' and number of objects pairs that are in different clusters under clustering C but in the same cluster under clustering C' remains nearly unchanged for 2, 3, 4 and 5 clusters. In case of Fowlkes - Mallows index and Jaccard index nearly in all cases indexes deteriorate with increasing the number of clusters. That means that number of objects pairs that are in the same cluster under clustering C but in different cluster under clustering C' and number of objects pairs that are in different clusters under clustering C but in the same cluster under clustering C' increases while number of objects that are in same classes in both clustering decreases. It can be concluded that the more clusters is used the more differences between clustering results we obtain. Comparing those two indexes with Rand index value it can concluded that there is a stable sum of objects clustered to the same cluster and clustered to different clusters while the more clusters we have the more objects pairs do not share the clustering ($n_{10}$ and $n_{01}$ values). All indexes indicated that Fuzzy and K-medians clustering gives very similar results as K-means approach while Hierarchical clustering and Expectation - maximization clustering results in smaller similarity. This situation might be cause by the fact that the data distribution might not be normal or by the presence of outliers [2]. Both situations happen very often due to presence of MoCap tracing inaccuracies that is commonly introduced to data acquired by multimedia systems that we used. Due to this the K-medians clustering was chosen to the second experiment as the fastest and most easily to interpreted method considering inequality (1) as the key-frame membership definition.

In the second experiment the average percentage of recordings in training and test set that contains all key frames detected by K-means algorithm is evaluated taking into account various number of clusters. As can be seen in Fig. 4 results in training dataset are very similar to results in test dataset. That means that key frames are well taught by our approach and the proposed method can deal with a new data. When the number of clusters is above 3 the percentage of detected key frames in recordings drops significantly. That means that best number of key frames that should be used in multimedia MoCap action classification systems is 2 or 3.

In the future research we plan to develop a fully automated method for appropriate features selection and key frames count determination for a given action type. We believe that it could be done with PCA which finds the features that have most contribution to variance or even PCA-derived variables can be used as features.

**Acknowledgments.** This work has been supported by the National Science Centre, Poland, under project number 2015/17/D/ST6/04051.

# References

1. Official website of gdl technology. http://gdl.org.pl. Access 24 Apr 2016
2. Almeida, J., Barbosa, L., Pais, A., Formosinho, S.: Improving hierarchical cluster analysis: a new method with outlier detection and automatic clustering. Chemometr. Intell. Lab. Syst. **87**(2), 208–217 (2007)
3. Bielecka, M., Piórkowski, A.: Automatized fuzzy evaluation of ct scan heart slices for creating 3d/4d heart model. Appl. Soft Comput. **30**, 179–189 (2015)
4. Ester, M., Kriegel, H.P., Sander, J., Xu, X., et al.: A density-based algorithm for discovering clusters in large spatial databases with noise. KDD **96**, 226–231 (1996)
5. Fraley, C., Raftery, A.E.: Bayesian regularization for normal mixture estimation and model-based clustering. J. Classif. **24**(2), 155–181 (2007)
6. Głowacz, A., Głowacz, W.: Diagnostics of dc machine based on sound recognition withapplication of FFT and jacquard distance. Przegląd Elektrotechniczny **86**, 292–295 (2010)
7. Hachaj, T., Ogiela, M.R.: Rule-based approach to recognizing human body poses and gestures in real time. Multimedia Syst. **20**(1), 81–99 (2014)
8. Hachaj, T., Ogiela, M.R.: Full body movements recognition-unsupervised learning approach with heuristic R-GDL method. Digit. Sig. Process. **46**, 239–252 (2015)
9. Hachaj, T., Ogiela, M.R., Koptyra, K.: Application of assistive computer vision methods to Oyama Karate techniques recognition. Symmetry **7**(4), 1670–1698 (2015)
10. Hachaj, T., Ogiela, M.R., Koptyra, K.: Human actions modelling and recognition in low-dimensional feature space. In: 2015 10th International Conference on Broadband and Wireless Computing, Communication and Applications (BWCCA), pp. 247–254. IEEE (2015)
11. Kaufman, L., Rousseeuw, P.J.: Finding Groups in Data: an Introduction to Cluster Analysis, vol. 344. Wiley, New York (2009)
12. Murtagh, F., Legendre, P.: Ward's hierarchical agglomerative clustering method: which algorithms implement ward's criterion? J. Classif. **31**(3), 274–295 (2014)
13. Wagner, S., Wagner, D.: Comparing clusterings: an overview. Universität Karlsruhe, Fakultät für Informatik Karlsruhe (2007)

# Image Similarity in Gaussian Mixture Model Based Image Retrieval

Maria Luszczkiewicz-Piatek[✉]

Faculty of Mathematics and Computer Science,
Department of Applied Computer Science, University of Lodz,
Banacha 22, 90-238 Lodz, Poland
mluszczkiewicz@math.uni.lodz.pl

**Abstract.** As color is a useful characteristic of our surrounding world, it gives clue for the recognition, indexing and retrieval of the images presenting the visual similarity. Thus, this paper focuses on the proper choice of the similarity measure applied to compare features evaluated in process the modeling of lossy coded color image information, based on the mixture approximation of chromaticity histogram. The analyzed similarity measure are those based on $Kullback - Leibler\ Diverence$, as $Goldberger\ approximation$ and $Variational\ approximation$. Signature-based distance function as $Hausdorff\ Distance$, $Perceptually\ Modified$ $Hausdorff\ Distance$ and $Earth\ Mover's\ Distance$ were also investigated. Retrieval results were obtained for $RGB, I1I2I3, YUV, CIE\ XYZ, CIE$ $L^*a^*b^*, HSx, LSLM$ and $TSL$ color spaces.

**Keywords:** Color image retrieval · Mixture modeling · Similarity measure · Color space · Lossy compression

## 1 Introduction

In recent years, the research on Content-based Image Retrieval (CBIR) has been very active, therefore developing and providing user-friendly tools for efficient image browsing and retrieval enables people utilizing their digital image collections. This huge amount of visual information is so far the largest and the most heterogenous image database, thus there is a question which features meaningfully describe its content. Color is a useful characteristic of our surrounding world, giving clue for the recognition, indexing and retrieval of the images presenting the visual similarity. However, the image representation in various color spaces can possibly yield different retrieval results due to the fact that employed color spaces can present different characteristics and thus they are suitable for different image processing tasks. Let us know that one of the common approaches is to represent each image as a feature vector and to assess similarity between images by a metric or similarity measure. Generative models such as $Gaussian$ $Mixture\ Model\ (GMM)$ are very suitable for that purposes due to their abilities to accurately approximate features of the analyzed images. Although the

© Springer International Publishing AG 2017
R.S. Choraś (ed.), *Image Processing and Communications Challenges 8*,
Advances in Intelligent Systems and Computing 525, DOI 10.1007/978-3-319-47274-4_10

choice of the color space plays a crucial role in image indexing and retrieval schemes, the question of which similarity measure provides the highest accuracy when comparing features of analyzed images, is significant.

In this paper the influence of the various similarity measures applied for comparison of features representing color images, on retrieval accuracy and efficiency is investigated. Thus, the presented paper is organized as follows. Firstly, the survey on color space set used for retrieval efficiency evaluation, is presented. Next Section introduces *Gaussian Mixture Model* applied for approximate color distribution within analyzed images. Then, the similarity measures applicable to *GMM*-based image retrieval methodology is presented along with the discussion on the experimental results. Finally, the conclusions are provided.

## 2   Analyzed Color Spaces

The color is uniquely defined in specified color space. The RGB color space is considered as fundamental and commonly used space, which is a base for many others obtained by it linear or nonlinear transformations. The color spaces evaluated by linear transformation of *RGB* (e.g. *YUV*, *YIQ*) are commonly associated with hardware color displays. On contrary, the color spaces obtained via nonlinear transformation of the *RGB* (e.g. *HSV* or *L*a*b***) are considered as reflection the characteristic of the human visual systems. Let us note that different color spaces derived from *RGB* by either group of linear or nonlinear transformation can reveal various performance. Thus it is important to determine which of the color spaces is the most desirable for image retrieval task.

First of the analyzed color spaces is the *XYZ* color space. It was derived by the International Commission on Illumination (*CIE*) in 1931 as a result of set of experiments on the human perception. The second analyzed color space $I_1I_2I_3$ is the result of the decorrelation of the RGB components using the *K-L* transform performed by Ohta, [1]. The next color space is the *YUV* color space used by color video standards. It consists of luminance ($Y$) component and chrominance components ($U$ and $V$). The *LSLM* is a color space based on the opposite responses of the cones, i.e. black-white, red-green and yellow-blue.

The nonlinear transformation of RGB color spaces are, among the others, *HSx*, *TSL* and *CIE L*a*b***. The *TSL* (Tint, Saturation, and Luminance) color space is widely used in face detection research field, [2]. *CIE L*a*b*** color space is a color-opponent space with dimension $L$ for lightness and $a$ and $b$ for the color-opponent dimensions, [3]. The *HSx* (Hue, Saturation) color spaces are related with the phenomenon of the human eye vision.

## 3   Gaussian Mixture Models for Image Modeling

The common application of the image retrieval techniques utilize the *Gaussian Mixture Model* (*GMM*) [4] as color distribution descriptor [5,6]. These methods simply index all images in the database by fitting GMM to the data, according to some predefined rules. The *mixture* based retrieval scheme applied in this work is robust to distortions introduced by lossy compression and noise, [7,8], thus it is important to test whether the use of various color spaces in information modelling process, and then the choice of similarity measure between evaluated models, will alternate the retrieval results, influencing the retrieval accuracy.

The first step in applying the applied methodology is to construct the histogram $H(x,y)$ in the chosen chromaticity space defined as $H(x,y) = N^{-1}\sharp\{r_{i,j} = x, g_{i,j} = y\}$, where $H(x,y)$ denotes a specified bin of a two-dimensional histogram with first component equal to $x$ and second component equal to $y$, the symbol $\sharp$ denotes the number of samples in a bin and $N$ is the number of color image pixels. As the pairs of the components representing analyzed color spaces the following pairs were chosen: $r-g$ (RGB), $I2I3$ (I1I2I3), $U-V$ (YUV), $T-S$ (TSL), $a-b$ ($L^*a^*b^*$), $H-S$ (HSx), and $S-LM$ (LSLM.)

The next stage of the presented technique is the modeling of the color histogram using the Gaussian Mixture Models (GMM) and utilizing the Expectation-Maximization (EM) algorithm for the model parameters estimation, [4]. Let us assume the following probabilistic model: $p(x|\Theta) = \sum_{m=1}^{M} \alpha_m p_m(x|\theta_m)$, which is composed of $M$ components and its parameters are defined as: $\Theta = (\alpha_1, \ldots \alpha_M, \theta_1, \ldots, \theta_M)$, with $\sum_{m=1}^{M} \alpha_m = 1$. Moreover, each $p_m$ is a function of the probability density function which is parameterized by $\theta_m$. Thus, the analyzed model consists of $M$ components with $M$ weighting coefficients $\alpha_m$. Finally after derivations the model parameters are defined as:

$$\alpha_m^{k+1} = N^{-1}\sum_{i=1}^{N} p(m|x_i, \Theta^k), \quad \mu_m^{k+1} = \frac{\sum_{i=1}^{N} x_i \cdot p(m|x_i, \Theta^k)}{\sum_{i=1}^{N} p(m|x_i, \Theta^k)}, \quad (1)$$

$$v_m^{k+1} = \frac{\sum_{i=1}^{N} p(m|x_i, \Theta^k)(x_i - \mu_m^{k+1})(x_i - \mu_m^{k+1})^T}{\sum_{i=1}^{N} p(m|x_i, \Theta^k)}, \quad (2)$$

where $\mu$ and $v$ denote the mean and variance, $m$ is the index of the model component and $k$ is the iteration number. The E (Expectation) and M (Maximization) steps are performed simultaneously, according to (1) and (2) and in each iteration, as the input data we use parameters obtained in the previous one.

The main idea of the application of the *GMM* technique lies in the highly desirable properties of this approach. The inherent feature of the GMM enables to approximate the distorted color histogram of the color image subjected to lossy compression, which is obtained through limited model complexity (7 components) and number of iterations (75) of E-M algorithm, as shown in [7,8]. Thus, this approach does not reflect exactly the corrupted (by e.g. lossy coding) data, but rather approximates it toward the chromaticity histogram of the original image. The lossy compression significantly corrupts the color distribution of an image and a lack of the application of any refinement techniques may lead to the high rate of false negative results, as images stored in lossy formats are considered as dissimilar on the basis of their corrupted color palette.

# 4   Comparison of Gaussian Mixture Models

Having image indexed using the GMM signatures, it is important to choose the accurate similarity measure to compare the indexes associated with each of the images with that of the given query. In general, the *Minkowski* metrics can be used to compare âĂİpoint to pointâĂİ the color histograms. However, these measures are very susceptible to bin shifts, thus even highly similar images can be considered as dissimilar when corresponding bins of their chromaticity histograms are shifted and therefore, this group of similarity measures is not taken into account in further analysis.

However, it is more suitable to generalize this concept toward âĂİdistribution to distributionâĂİ similarity. For that purpose the *Kullback-Leibler* (*K-L*) based dissimilarity measures as ($D_{KL}$) [9] can be applied, which can be solved analytically for normal distributions and that is taking into account the correlation of components within each of the both mixture models.

The following formula describe the $D_{KL}$ measure as follows:

$$D_{KL}(G_i, G_j) = (\mu_i - \mu_j)^T \Sigma^{-1}(\mu_i - \mu_j) + TR(\Sigma_i \Sigma_j^{-1} + \Sigma_i^{-1}\Sigma_j) \qquad (3)$$

where $G_i$ and $G_j$ denotes normal distributions with mean values $\mu_i$ and $\mu_j$, and covariance matrices $\Sigma_i$ and $\Sigma_j$ respectively.

In recent years there were proposed multiple approaches to approximation to of the *K-L* Divergence between Gaussian mixture models, i.e. *Goldberger* [9] approximation. This approach is based on the idea to match each mixture component of the fist GMM to one single component of the second GMM. The approximation is dependent on the *K-L* Divergence between the mixture components, as described by following formula:

$$D_G(G^a, G^b) = \sum_{i=1}^{k} \alpha_i^a (D_{KL}(G_i^a, G_{\pi(i)}^b) + \log \frac{\alpha_i^a}{\alpha_{\pi(i)}^b}) \qquad (4)$$

where $\pi(i) = \arg\min_j(D_{KL}(G_i^a, G_j^b) - \log \alpha_j^b$ is the matching function between components i and j of the Gaussian Mixture Models $G^a$ and $G^b$ and $\alpha_i^b$ denotes the weighting parameter $\alpha$ for $i^{th}$ component of model $G^a$. Let us note that in fact *Goldberger* approximation is sum of K-L Divergences $D_{KL}(G_i^a, G_{\pi(i)}^b)$ for matching mixture components $G_i^a, G_{\pi(i)}^b$ and logarithm of the ratio of respective prior probabilities $\log \frac{\alpha_i^a}{\alpha_{\pi(i)}^b}$. Although this measure can be considered as very suitable for mixture model based color image retrieval there are known drawbacks of this approach [10, 11]. Firstly, this method fails in case of components which are overlapping. Let us underline, that the flexibility of mixture model to approximate the color information of the image, is provided by possible overlap of components with adjustable weights, and it is one of the main reasons to use $GMM$ approach for modelling of color image information content. Secondly this method produce questionable results in case of low component weights. These two significant problems lead to the conclusion that *Goldberger* approximation, if used for color image retrieval based on GMM approach, should be used with great care.

Next measure, the *Variational approximation* $D_V$ [10] is defined as:

$$D_V(G^a, G^b) = \sum_{i=1}^{k} \alpha_i^a \frac{\sum_{i'=1}^{k} \alpha_{i'}^a e^{-D_{KL}(G_i^a, G_{i'}^b)}}{\sum_{j=1}^{k} \alpha_j^b e^{-D_{KL}(G_i^a, G_j^b)}} \tag{5}$$

$D_V$ differs from aforementioned method because it takes into account $D_{KL}$ of mixture components of $G_i^a$ and also $D_{KL}$ between both models $G_i^a$ and $G_j^b$. Let us note that $D_V$ approach was initially designed for acoustic models used for speech recognition.

To summarize, it can be noticed that presented measures differ in scheme of incorporating components for both GMM (i.e. $G^a$ and $G^b$) into similarity calculations.

The second group of analyzed similarity measures are those which take into account all information conveyed by each component of both GMM in form of signature based measures. These signatures can be either feature histograms or feature signatures. As a measure of the distance between two distributions in these frameworks the K-L Divergence was used, due to its ability to incorporate full information (mean and covariance) conveyed by each component of mixture.

One of such approaches is *Earth Movers Distance* [12] (*EMD*), which is based on the assumption that one of the histograms reflects hills and the second represents holes in the ground of a histogram. The measured distance is defined as a minimum amount of work needed to transform one histogram into the other using a soil of the first histogram.

The *Hausdorf Distance* ($D_H$) [13] measures the maximum nearest neighbour distance, i.e. K-L Divergence, among components of both GMM ($G^a$ and $G^b$):

$$D_H(G^a, G^b) = \max\{h(G^a, G^b), h(G^b, G^a)\} \tag{6}$$

where $h(G^a, G^b) = \max_{G^a} \min_{G^b} D_{KL}(G^a, G^b)$

For this method two GMMs are regarded as not distant if corresponding components of both models have similar parameter values. Let us note that this approach does not take into account the weighting parameters $\alpha$, recognizing as similar GMMs approximating histograms of different structures, as weighting parameters $\alpha$, control the shape of the histogram approximation for models of the Gaussian distributions of the same parameters $\mu$ and $\Sigma$.

The modified version of *Hausdorf Distance* takes into account weighting parameters $\alpha$ in form of modification of auxiliary function $h(G^a, G^b)$. Thus, the *Perceptually Modified Hausdorf Distance* ($D_{PMH}$) is defined as, [14]:

$$D_{PMH}(G^a, G^b) = \max\{h_m(G^a, G^b), h_m(G^b, G^a)\} \tag{7}$$

where $h_m(G^a, G^b) = \frac{\sum_{i=1}^{k} \alpha_i^a \cdot \min_{G^b} \frac{D_{KL}(G^a, G^b)}{\min(\alpha^a, \alpha^b)}}{\sum_{i=1}^{k} \alpha_i^a}$

$D_{PMH}$ diminish an influence of components of low importance i.e. those with low values of weighting parameter $\alpha$.

The last of the analyzed approaches to determine the similarity between GMMs is *Bhattaharyya-based distance* $(D_B)$ [15].

$$D_B(G^a, G^b) = \sum_{i=1}^{k} \sum_{j=1}^{m} \alpha_i^a \cdot \alpha_j^b \cdot B(G_i^a, G_j^b) \tag{8}$$

where $B(G_i^a, G_j^b) = \frac{1}{8}(\mu_i^a - \mu_j^b)^T (\frac{\Sigma_i^a + \Sigma_j^b}{2})^{-1}(\mu_i^a - \mu_j^b) + \frac{1}{2}\ln\frac{|\frac{\Sigma_i^a + \Sigma_j^b}{2}|}{\sqrt{|\Sigma_i^a| \cdot |\Sigma_j^b|}}$.

## 5   Experimental Results

In order to choose the most accurate retrieval criteria the set of eight color spaces ($RGB$, $CIE\ XYZ$, $YUV$, $I1I2I3$, $LSLM$, $CIE\ L^*a^*b^*$, $HSx$ and $TSL$) and six similarity measures ($EMD$, $D_G$, $D_V$, $D_H$, $D_{PMH}$, $D_B$) were analyzed. Firstly, the color images of database of Wang [16] (1000 color images categorized into 10 thematically consistent categories), were compressed to 25 % of their file size and rescaled to 10 % of their size, using IrfanView software. These images produced distorted histograms in comparison to the histograms of the original, uncompressed images. Next step was to model the color chromaticity histograms using the $GMM$ methodology obtaining image signatures, according to formulae presented in previous Sections. These image signatures were then compared by the set of mentioned similarity measures. To generalize retrieval observations, the *Precision* and *Recall* measures are employed. In more details, *Precision* is the fraction of retrieved images that are relevant, while Recall is the fraction of relevant instances that are retrieved.

The criterion of the image similarity was the membership of the image to one of the 10 categories of this database.

In this work the efficiency of the analyzed retrieval schemes was not shown on *Precision* − *Recall* plots. Let us note, that *Precision* and *Recall* tend to ignore the ranking of retrieved images, e.g. *Precision* of 0.5 indicates that half of the retrieved images is relevant to the given query but without any further information if relevant images are the first half of the retrieved set or are the second half. In order to counteract this, it is advised to measure *Precision* at the specified points e.g. at the answer set of 1, 5, 10, and 25 images. These measurements are assumed to give the better idea of the behavior of the analyzed retrieval scheme, because user is usually interested only in limited number of retrieved images indicated as most relevant to given query. The other interesting measure of the overall quality and accuracy of the retrieval is *Mean Average Precision* which is mean of the average precision scores for each query.

It can be noticed that the $CIE\ L*a*b*$ (nonlinear), $RGB$ and $I1I2I3$ (linear) color spaces offer the best retrieval efficiency. One must be aware that the retrieval efficiency is not only related with the retrieval method, but also with the content of the analyzed database and the used ground truth. In the case of the Wang database some of the categories share not only the semantic relation between images within the category, but also the arrangement of the colors

**Table 1.** The Average precision $(\hat{P})$ values of the retrieval results evaluated using GMM methodology for set of color spaces. Average precision values at specified points related to 1, 5, 10 and 25 retrieved images were obtained over all the images of the database of Wang

| Color space | $\hat{P}_1$ | $\hat{P}_5$ | $\hat{P}_{10}$ | $\hat{P}_{25}$ | $\hat{P}_1$ | $\hat{P}_5$ | $\hat{P}_{10}$ | $\hat{P}_{25}$ | $\hat{P}_1$ | $\hat{P}_5$ | $\hat{P}_{10}$ | $\hat{P}_{25}$ |
|---|---|---|---|---|---|---|---|---|---|---|---|---|
| | EMD | | | | $D_G$ | | | | $D_V$ | | | |
| RGB | 1 | 0.6052 | 0.5176 | 0.4368 | 0.3384 | 0.1659 | 0.1522 | 0.1174 | 0.1724 | 0.1077 | 0.1039 | 0.1250 |
| XYZ | 1 | 0.6225 | 0.5365 | 0.4429 | 0.3630 | 0.1963 | 0.1765 | 0.1757 | 0.1918 | 0.1207 | 0.1232 | 0.1371 |
| YUV | 1 | 0.6240 | 0.5505 | 0.4587 | 0.3929 | 0.2021 | 0.1826 | 0.1786 | 0.1923 | 0.1130 | 0.1079 | 0.1101 |
| I1I2I3 | 1 | 0.6491 | 0.5833 | 0.4750 | 0.3440 | 0.1944 | 0.1798 | 0.1784 | 0.1526 | 0.0989 | 0.1012 | 0.1122 |
| LSLM | 1 | 0.5908 | 0.4925 | 0.3781 | 0.3984 | 0.1923 | 0.1674 | 0.1745 | 0.1584 | 0.1039 | 0.1063 | 0.1179 |
| $L^*a^*b^*$ | 1 | 0.6491 | 0.5833 | 0.4750 | 0.3958 | 0.1848 | 0.1757 | 0.1528 | 0.1485 | 0.0956 | 0.0995 | 0.1086 |
| HSx | 1 | 0.6283 | 0.5559 | 0.4722 | 0.3766 | 0.1360 | 0.1204 | 0.1198 | 0.1845 | 0.1123 | 0.1143 | 0.1239 |
| TSL | 1 | 0.6374 | 0.5524 | 0.4462 | 0.3696 | 0.13675 | 0.1285 | 0.1193 | 0.1066 | 0.1254 | 0.1252 | 0.1323 |
| | $D_H$ | | | | $D_{PMH}$ | | | | $D_B$ | | | |
| RGB | 1 | 0.4055 | 0.2792 | 0.1736 | 0.5010 | 0.3913 | 0.3379 | 0.2370 | 0.4530 | 0.3554 | 0.3157 | 0.2985 |
| XYZ | 1 | 0.4936 | 0.4045 | 0.2825 | 0.8122 | 0.4919 | 0.4377 | 0.3613 | 0.4083 | 0.3486 | 0.3324 | 0.3155 |
| YUV | 1 | 0.3657 | 0.2892 | 0.2220 | 0.7206 | 0.4123 | 0.3528 | 0.2845 | 0.3194 | 0.2325 | 0.2317 | 0.2128 |
| I1I2I3 | 1 | 0.4583 | 0.3774 | 0.2870 | 0.8241 | 0.5036 | 0.4431 | 0.3520 | 0.3428 | 0.2420 | 0.2378 | 0.2266 |
| LSLM | 1 | 0.4853 | 0.3642 | 0.2578 | 0.5783 | 0.4103 | 0.3729 | 0.3098 | 0.3214 | 0.2438 | 0.2122 | 0.1939 |
| $L^*a^*b^*$ | 1 | 0.4505 | 0.3559 | 0.2750 | 0.8441 | 0.5070 | 0.4412 | 0.3575 | 0.3675 | 0.2781 | 0.2594 | 0.2378 |
| HSx | 1 | 0.4130 | 0.3296 | 0.2399 | 0.6121 | 0.4284 | 0.3647 | 0.2802 | 0.3748 | 0.2844 | 0.2624 | 0.2279 |
| TSL | 1 | 0.4380 | 0.3433 | 0.2569 | 0.8778 | 0.5022 | 0.4243 | 0.3192 | 0.2792 | 0.2449 | 0.2421 | 0.2324 |

**Table 2.** The Mean Average Precision $(MAP)$ values of the retrieval results evaluated using GMM methodology for set of color spaces. $MAP$ values were obtained for all of the images of the database of Wang

| Color space | EMD | $D_G$ | $D_V$ | $D_H$ | $D_{PMH}$ | $D_B$ |
|---|---|---|---|---|---|---|
| RGB | 0.3791 | 0.1267 | 0.1197 | 0.1821 | 0.200 | 0.2652 |
| XYZ | 0.3873 | 0.1406 | 0.1373 | 0.2432 | 0.3022 | 0.2909 |
| YUV | 0.3820 | 0.1487 | 0.1137 | 0.2139 | 0.2441 | 0.2109 |
| I1I2I3 | 0.3943 | 0.1442 | 0.1153 | 0.2439 | 0.2871 | 0.2305 |
| LSLM | 0.3231 | 0.1365 | 0.1180 | 0.2360 | 0.2508 | 0.1982 |
| $L^*a^*b^*$ | 0.4052 | 0.1453 | 0.1124 | 0.2417 | 0.2914 | 0.2450 |
| HSx | 0.3837 | 0.1220 | 0.1245 | 0.2149 | 0.2368 | 0.1991 |
| TSL | 0.3576 | 0.1253 | 0.1304 | 0.2299 | 0.2679 | 0.2202 |

present on the images, what influences the retrieval efficiency of applied technique. It is important to underline that this work does not test the effectiveness of the $GMM$ based retrieval scheme, as it efficiency was elaborated and tested in other Author's work i.e. [7,8] as well as the comparison with other widely used retrieval schemes as $MPEG-7$ descriptors, correlograms and others. The previous work clearly show the usefullness of the $GMM$ based retrieval technique, especially when lossy compressed images are analyzed. Presented paper addresses the problem which similarity measure between GMMs and which color space is best suited for this kind of retrieval scheme. Table 1 summarizes the average precision for the entire database of Wang at the points of 1, 5, 10 and 25 retrieved images for six previously discussed similarity measures. It can be observed that $L*a*b*$ and $I1I2I3$ color spaces provide comparable results for small set of retrieved candidate images. It can be observed that $EMD$ measure yield best retrieval results. The $D_G$ and $D_V$ provided the worst results and thus they are not well suited for image modeling and retrieval based on mixture approach. The average precision values at the points related to 1, 5, 10 and 25 retrieved images should be chosen to examine the retrieval efficiency as user is usually more interested in relevance of the highly ranked candidate images than the overall success rate of the retrieval system.

Table 2 summarizes the Mean Average Precision evaluated for the entire database indicating that $L*a*b*$ color space and $EMD$ similarity measure are the most desirable for mixture based retrieval. $D_{PMH}$ is theoretically better suited for mixture-based image retrieval and it provides better efficiency for higher number of retrieved images for a given query than $D_H$, but it provides lower accuracy when number of retrieved images is smaller.

## 6    Conclusions

In this work the problem of the choice of the most accurate similarity measure and color space for the $GMM$ based retrieval scheme was analyzed. The conducted experiments shown that this decision plays a crucial role in the efficiency of the retrieval system. Thus, the best performance of the $GMM$ based scheme is associated with the $EMD$ similarity measure using the CIE $L^*a^*b^*$ color spaces. As CIE $L^*a^*b^*$ slightly outperforms the $I1I2I3$ it should be taken into account that when image color information is given using $RGB$ values, the linear transformation to $I1I2I3$ is less complicated than for CIE $L^*a^*b^*$ Due to the fact that user is, in general, interested in relatively small set of top ranked images, it is important to examine the system efficiency at points of e.g. 1, 5 and 10 retrieved images. This comparison (see Table 1) also indicates that CIE $L^*a^*b^*$ and $I1I2I3$ color spaces are the most accurate choice. However, Table 2 examines the retrieval efficiency in terms of *Mean Average Precision* evaluated for all images in analyzed database. This experiment shows that CIE $L^*a^*b^*$ color space outperforms all analyzed color spaces.

**Acknowledgment.** This work has been supported by The National Science Centre under SONATA grant no. UMO-2011/01/D/ST6/04554.

# References

1. Ohta, Y., Kanade, T., Sakai, T.: Color information for region segmentation. Comput. Graph. Image Processing. **13**(3), 222–241 (1980)
2. Hsu, R.-L., Abdel-Mottaleb, M., Jain, A.K.: Face detection in color images. IEEE Trans. Pattern Anal. Mach. Intell. **24**(5), 696–706 (2002)
3. Fairchild, M.D.: Color and Image Appearance Models, Color Appearance Models. Wiley, New York (2005)
4. McLachlan, G., Peel, D.: Finite Mixtures Models. Wiley, New York (2000)
5. Jeong, S., Won, C.S., Gray, R.M.: Image retrieval using color histograms generated by gauss mixture vector quantization. Comp. Vis. Image Underst. **94**(1–3), 44–66 (2004)
6. Xing, X., Zhang, Y., Gong, B.: Mixture model based contextual image retrieval. In: CIVR 2010, pp. 251–258 (2010)
7. Luszczkiewicz, M., Smolka, B.: Application of bilateral filtering and Gaussian mixture modeling for the retrieval of paintings. In: Proceedings of 16th IEEE International Conference on Image Processing (ICIP), pp. 77–80 (2009)
8. Luszczkiewicz-Piatek, M., Smolka, B.: Selective color image retrieval based on the gaussian mixture model. In: Blanc-Talon, J., Philips, W., Popescu, D., Scheunders, P., Zemčík, P. (eds.) ACIVS 2012. LNCS, vol. 7517, pp. 431–443. Springer, Heidelberg (2012). doi:10.1007/978-3-642-33140-4_38
9. Goldberger, J., et al.: An efficient image similarity measure based on approximation of KL divergence between two gaussian mixtures. In: ICCV, pp. 487–493 (2003)
10. Harshey, J., Olsen, P.: Approximating the kullback-leibler divergence between gaussian mixure models. In: Proceedings of IEEE ICASSP, vol. 4, pp. 317–320 (2007)
11. Huo, Q., Li, W.: A DTW-based dissimilarity measure for left-to-right hidden Markov models and its application to word confusability analysis. In: INTER-SPEECH, pp. 2338–2341 (2006)
12. Rubner, Y., Tomasi, C., Guibas, L.J.: The earth mover's distance as a metric for image retrieval. Int. J. Comput. Vis. **40**(2), 99–121 (2000)
13. Huttenlocher, D.P., Klanderman, G.A., Rucklidge, W.J.: Comparing images using the Hausdorff distance. IEEE Trans. Pattern Analy. Mach. Intell. **15**(9), 850–863 (1993)
14. Park, B.G., Lee, K.M., Lee, S.U.: Color-based image retrieval using perceptually modified hausdorff distance. EURASIP J. Image Video Proces. **4**, 1–10 (2008)
15. Fukunaga, K.: Introduction to Statistical Pattern Recognition. Academic Press, London (1990)
16. Wang, J.Z., Li, J., Wiederhold, G.: SIMPLIcity: semantics - sensitive integrated matching for picture libraries. IEEE Trans. Pattern Anal. Mach. Intel. **9**, 947–963 (2001)

# A Flexible Software Architecture for a Network of Heterogeneous Smart Cameras

Dominik Pieczyński, Marek Kraft[(✉)], and Michał Fularz

Institute of Control and Information Engineering, Poznań University of Technology,
Piotrowo 3a, 60-965 Poznań, Poland
marek.kraft@put.poznan.pl

**Abstract.** This paper presents a flexible software solution, facilitating easy deployment and control of individual sensor nodes in a camera network. The presented solution is lightweight, extensible and provides an easy to use base for the implementation of a range of collaborative tasks performed by multi-camera setups.

## 1 Introduction

In recent years, video surveillance has become a widespread technology. With the constant growth of demand for visual supervision, increasing camera and computational platform capability and their price decrease one can predict, that the growth in this sector will be sustained for year to come. As the scale of video surveillance systems increases and the number of cameras in a typical surveillance system grows larger, the analysis of data by human operators becomes increasingly hard, or even impossible because of the sheer scope and volume of visual data. Moreover, scientific studies reveal, that tiredness and loss of attention over prolonged periods of time is a serious issue with human operators of video surveillance systems [8]. The most common way to remedy those issues is the employment of automated, image processing based surveillance.

Automated surveillance is, however, not without its own issues. Extraction of useful data from surveillance camera images is not a trivial task. Most image and video processing algorithms are considered computationally intensive, which is an even bigger concern in the case of large scale systems. Moreover, to make use of the full potential of a number of cameras, one should not treat them as isolated sensors and try to employ a holistic approach to sensor data analysis. A straightforward approach to data aggregation is to use a centralized server for all the processing. This, however, usually calls for constant transmission of video streams, resulting in a significant strain on the communication infrastructure. In extreme cases, the communication channels may not be able to support continuous transmission from all the installed cameras. One way to remedy this is the use of constantly improving video compression methods. This approach,

The original version of this chapter was revised: The spelling of the first author's name was corrected. The erratum to this chapter is available at DOI: 10.1007/978-3-319-47274-4_34

R.S. Choraś (ed.), *Image Processing and Communications Challenges 8*,
Advances in Intelligent Systems and Computing 525, DOI 10.1007/978-3-319-47274-4_11

however, is not without its own issues. Video compression requires additional computational power and contributes to the power consumption of the cameras in the network [11]. Moreover, increasing the compression rate may have a negative impact on the results of the image processing algorithms applied to the transmitted images [7]. The limitations of the central processing paradigm can be overcome by using distributed processing.

In the case of distributed processing, a significant portion of computations is performed by the camera network nodes themselves, following the 'smart camera' concept. Since modern embedded processors are currently powerful enough to handle typical workloads associated with real-time image and video processing, the main limitation of such an approach becomes less and less significant [3, 5]. Nodes of such camera networks are to a large extent autonomous, and can communicate with the central server and each other [1,4]. With the majority of image processing operations performed in-place, the amount of transmitted data can be significantly reduced. Taking full advantage of network-wide information raises the need for specialized algorithms for node and network management and control. Moreover, the flexibility and scalability of software that manages such vast networks is a very desirable feature.

In this paper, we describe an architecture of a software system capable of governing a network of heterogeneous smart cameras. The network performs the task of adaptive activity monitoring [10], but can be easily adapted to host a range of other image and video processing applications. Smart cameras of different types and architectures can dynamically join and leave the network. Moreover, they are being made aware of their mutual relations and receive software updates automatically. Ethernet or WiFi wireless network can be used for communication, making the integration with existing infrastructure virtually effortless.

## 2  Related Work

While the methods for data processing and the communication technology utilised in camera networks show constant progress, less attention is paid to system-level software (middleware). The focus is mainly on the design of individual sensor nodes facilitating node communication using a variety of protocols [6,13] or information fusion [2] and collaborative processing in a variety of scenarios [15]. This, however, is bound to change, as proper middleware makes the design, implementation and deployment of vast camera network significantly less troublesome. In [14] the authors propose a camera network software architecture which moves the majority of processing to a cloud-based solution. According to the authors, the smart cameras are not well suited for high-performance image processing, hence the data is moved, aggregated and processed in virtual local hubs. The closest counterpart to the solution described in this was presented in [9]. It is also an architecture and system independent solution, but based on a more general publisher/subscriber architecture and its potential use cases extend beyond camera networks.

## 3   Description of the Implemented System

The main contribution of this paper is a software framework that facilitates the deployment and testing of new solutions for visual sensor networks. It consist of two separate parts - one is a firmware for a single smart camera (sensor node) while the other one is a network coordinator module.

The described middleware offers a few distinctive advantages:

- automatic discovery of new nodes that join the network,
- centralized way of distributing the messages,
- general architecture for a software dedicated for a smart camera,
- ability to automatically update all the nodes when the new firmware version is deployed on the server,
- support for different types of hardware, as the software is hardware-agnostic,
- ability to remotely set up the parameters of a camera,
- an optional UI on the server, enabling system monitoring.

The block schematic of the proposed system is shown in the Fig. 1.

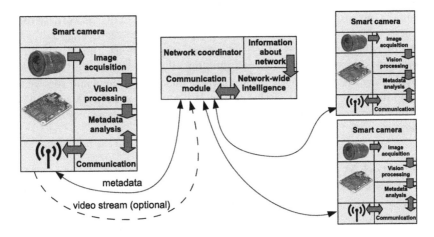

**Fig. 1.** Network schematic

The smart camera software is split into four cooperating and interchangeable modules, responsible for following tasks: image acquisition, vision processing, metadata analysis and communication. Partitioning tasks in such way makes it really easy to replace one of the modules. The image acquisition module captures the data from the image sensor and configures low level parameters of the camera. In our example code two different implementations are provided - one for dedicated Raspberry Pi camera using CSI (Camera Serial Interface) and the other one using general UVC (USB video class). The vision processing module is supposed to extract the high level information from the acquired images.

It is tightly integrated with the established OpenCV video processing library, enabling quick and efficient implementation of desired functionality. Implementation of average median algorithm for background subtraction and movement detection (Sect. 4) is provided with the code as an example. After the image processing step and extraction of high level information about the scene, the metadata analysis module is charged with the task of interpreting this information and taking whatever action is appropriate like eg. sending an alarm message to others nodes of the system, performing additional image analysis etc. The final module is responsible for communication with the rest of the system. It is able to update the camera's other modules when the new version of the system software is available. It is also responsible for the procedure of joining the network and negotiating the node's parameters.

The software for the network coordinator is also modular. The communication module is similar to the one found in the smart camera – it is responsible for managing data transfers with other nodes of the network. The network-wide intelligence module is gathering the information from all the cameras and, using stored information about the network topology, is capable of performing actions like sending information to all or just a few nodes of the network or controlling what algorithms are executed in each node. In the example application, the coordinator gathers data about the activity (defined as the amount of movement in the scene observed by the camera) from all of the nodes and resends that data augmented with the information about the activity of the neighbours [10].

The described framework is written using high level Python programming language in a way to make it as hardware agnostic as possible. As an example three different processing platforms were used to test the solution. The requirements are modest and limited to the operating system capable of running Docker application container engine. In addition to that the developed software relies on open sources solutions like Linux operating system and an established image processing library (OpenCV). All this makes the software accessible and easy to use by other researchers, so that they can deploy and test new solutions for visual sensor networks.

## 4   Image Processing

The activity on the scene observed by any given camera in the network is computed based on a background subtraction algorithm. Approximate median algorithm presented in [12] was selected for its simplicity and relatively good quality of results. The background model is updated by performing the following steps:

- if the intensity value $I_{x,y}$ of currently investigated pixel of the current frame $I$ is greater than the value of the corresponding background model pixel $B_{x,y}$, the value of $B_{x,y}$ is incremented by one,
- if the intensity value $I_{x,y}$ of currently investigated pixel of the current frame $I$ is smaller than the value of the corresponding background model pixel $B_{x,y}$, the value of $B_{x,y}$ is decremented by one,

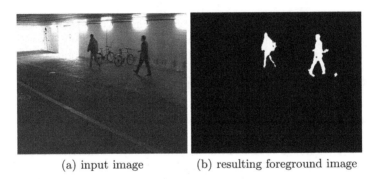

(a) input image          (b) resulting foreground image

**Fig. 2.** Example results from the background subtraction pipeline. Moving (active) objects are successfully detected and highlighted

– if the intensity value $I_{x,y}$ of currently investigated pixel of the current frame $I$ has the same value as the corresponding background model pixel $B_{x,y}$, the value of $B_{x,y}$ remains unchanged.

Computing the difference image $D$ as an absolute value of the difference between the current frame $I$ and the background model $B$ returns the current foreground. The difference image is subjected to additional post-processing using a $3 \times 3$ Gaussian, fixed-threshold binarization and binary morphological closing to improve the quality of results and foreground consistency. The percentage of foreground pixels reflects the activity observed by the local sensor node. Example output produced by the background subtraction pipeline is shown in Fig. 2.

## 5   Communication Scheme

TCP, abstracted by the Transports and Protocols API from Python's asyncio module, is used as the transport protocol. Binary packets, encoded using MessagePack module are sent in a bidirectional fashion between clients and a server.

The server, when started, opens a port, which the cameras will later connect to. Additionally, the server begins the process of discovering Docker daemons available in the network using mDNS/DNS-SD protocols. When a new device is discovered, the server connects to the client's Docker daemon, downloads or updates the service image if necessary and starts a new application container. The server's address is passed to the container during creation as an environment variable, allowing the client software to begin negotiating settings with the server. When the server detects a new connection from the client software, it sends a configuration packet to that device. After accepting new settings the client sends the packet containing its identifier, based on the device's hostname, back to the central node. The client device becomes acknowledged by the server and can start transmitting data packets.

The central node processes all the packets from the connected clients and calculates camera's neighbours activity level. The result is then transmitted to the appropriate devices.

There is also a possibility that the change of camera settings is needed. In such case, the server software prepares a new configuration packet and fills it with the requested settings' values. This packet is then transmitted to the client, which changes the settings accordingly.

The connection termination is normally initialized by the server, which performs a standard TCP Connection Termination.

## 6    Results and Discussion

The prepared solution was tested on the Raspberry Pi model B evaluation board (with Broadcom BCM2835 processor - ARMv6 single core, 0.7 GHz), Odroid XU4 evaluation board (with Samsung Exynos 5422 processor - ARMv7 octa core, 2 GHz) and notebook computer (with AMD E2-1800 processors - x86-64 compatible dual core, 1.7 GHz).

The same application was run on each of the targets and the CPU usage and data transfer was recorded. The application analyses the observed scene and detects the movement. If the amount of movement detected in a node and

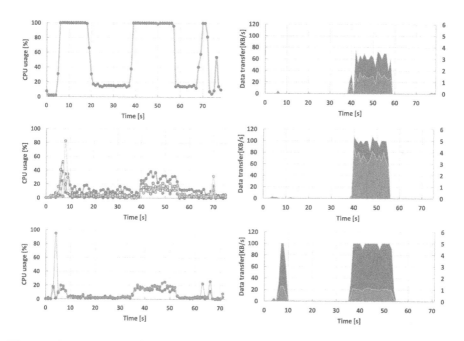

**Fig. 3.** The CPU usage (on the left) and the amount of data sent and received (on the right). First row shows results for Raspberry Pi, second one for the Odroid, while the last one shows results for PC. On the CPU usage charts different colors denote the usage of different CPU cores (1 for RPi, 4 for Odroid and 2 for PC). Blue color on the right charts represents the amount of data sent with scale on left axis; the orange color represents the amount of data received with scale on the right axis (Color figure online)

received from neighbouring nodes exceeds the threshold, the smart camera starts sending the images to the network coordinator for recording. The gathered data is shown if Fig. 3.

The figures shows that the usage of processor and amount of data transferred is high only during the periods of activity, when system processes 15 frames per second and sends them to server. During the normal operation, the processor usage is modest and the amount of data transferred is negligible, which shows that the proposed middleware software does not impose a significant load on the processor or the communication interface. It is clearly visible, that the slowest target – the Raspberry Pi – is not capable of achieving the 15 fps, which influences the amount of data sent.

## 7   Conclusions

The paper presents a scalable, flexible software, serving as a base for operation of a network of distributed smart camera sensors. The described software facilitates the deployment, configuration and testing of remote sensor nodes, which greatly simplifies the development of applications involving multi-camera systems. Moreover, the solution's computational complexity is very low, enabling its use on low-end hardware without a significant overhead. The software is released with a permissive open source license and can be downloaded from GitHub[1,2].

**Acknowledgement.** This research was financed by the Polish National Science Centre grant funded according to the decision DEC-2011/03/N/ST6/03022, which is gratefully acknowledged.

## References

1. Aghajan, H., Cavallaro, A.: Multi-camera Networks: Principles and Applications. Academic press, London (2009)
2. Bajo, J., Paz, J.F.D., Villarrubia, G., Corchado, J.M.: Self-organizing architecture for information fusion in distributed sensor networks. Int. J. Distrib. Sen. Netw. **2015**, 2 (2015)
3. Belbachir, A.N.: Smart Cameras. Springer, Heidelberg (2010)
4. Bhanu, B., Ravishankar, C., Roy-Chowdhury, A., Aghajan, H., Terzopoulos, D.: Distributed Video Sensor Networks. Springer, London (2011)
5. Bobda, C., Velipasalar, S.: Distributed Embedded Smart Cameras. Springer, New York (2014)
6. Chen, P., Hong, K., Naikal, N., Sastry, S.S., Tygar, D., Yan, P., Yang, A.Y., Chang, L.C., Lin, L., Wang, S., Lobatón, E., Oh, S., Ahammad, P.: A low-bandwidth camera sensor platform with applications in smart camera networks. ACM Trans. Sens. Netw. **9**(2), 21:1–21:23 (2013)

---

[1] https://github.com/PUTvision/VSNServer.
[2] https://github.com/PUTvision/VSNClient.

7. Cozzolino, A., Flammini, F., Galli, V., Lamberti, M., Poggi, G., Pragliola, C.: Evaluating the effects of MJPEG compression on motion tracking in metro railway surveillance. In: Blanc-Talon, J., Philips, W., Popescu, D., Scheunders, P., Zemčík, P. (eds.) ACIVS 2012. LNCS, vol. 7517, pp. 142–154. Springer, Heidelberg (2012). doi:10.1007/978-3-642-33140-4_13

8. Dadashi, N., Stedmon, A., Pridmore, T.: Semi-automated CCTV surveillance: the effects of system confidence, system accuracy and task complexity on operator vigilance, reliance and workload. Appl. Ergon. 44(5), 730–738 (2013)

9. Dieber, B., Simonjan, J., Esterle, L., Rinner, B., Nebehay, G., Pflugfelder, R., Fernandez, G.J.: Ella: middleware for multi-camera surveillance in heterogeneous visual sensor networks. In: 2013 Seventh International Conference on Distributed Smart Cameras (ICDSC), pp. 1–6. IEEE (2013)

10. Kraft, M., Fularz, M., Schmidt, A.: Collaborative, context based activity control method for camera networks. In: Battiato, S., Blanc-Talon, J., Gallo, G., Philips, W., Popescu, D., Scheunders, P. (eds.) ACIVS 2015. LNCS, vol. 9386, pp. 118–129. Springer, Heidelberg (2015). doi:10.1007/978-3-319-25903-1_11

11. Ma, T., Hempel, M., Peng, D., Sharif, H.: A survey of energy-efficient compression and communication techniques for multimedia in resource constrained systems. Commun. Surv. Tutor. IEEE 15(3), 963–972 (2013)

12. McFarlane, N., Schofield, C.: Segmentation and tracking of piglets in images. Mach. Vis. Appl. 8, 187–193 (1995)

13. Miller, L., Abas, K., Obraczka, K.: Scmesh: solar-powered wireless smart camera mesh network. In: 2015 24th International Conference on Computer Communication and Networks (ICCCN), pp. 1–8, August 2015

14. Saini, M.K., Atrey, P.K., Saddik, A.E.: From smart camera to smarthub: embracing cloud for video surveillance. Int. J. Distrib. Sens. Netw. 2014, 1–10 (2014)

15. Tessens, L., Morbee, M., Aghajan, H., Philips, W.: Camera selection for tracking in distributed smart camera networks. ACM Trans. Sen. Netw. 10(2), 23:1–23:33 (2014)

# Quality Assessment of 3D Prints Based on Feature Similarity Metrics

Krzysztof Okarma$^{(\boxtimes)}$ and Jarosław Fastowicz

Faculty of Electrical Engineering,
Department of Signal Processing and Multimedia Engineering,
West Pomeranian University of Technology, Szczecin,
26. Kwietnia 10, 71-126 Szczecin, Poland
{krzysztof.okarma,jaroslaw.fastowicz}@zut.edu.pl

**Abstract.** Visual quality inspection of 3D prints is one of the most recent challenges in image quality assessment domain. One of the natural approaches to this issue seems to be the use of some existing metrics successfully applied to general image quality assessment purposes. Since the application of basic Structural Similarity does not lead to satisfactory quality prediction of 3D prints, in this paper some experimental results obtained using feature based metrics have been presented. Due to the use of different colors of filaments the influence of color to grayscale conversion method has also been analyzed. Proposed approach leads to promising results allowing a reliable prediction of 3D prints quality for different colors of filaments.

## 1 Introduction

The 3D printing is currently one of the most dynamically growing technology utilized to quickly create physical objects and prototypes based on virtual 3D models. This field is rapidly developing in all fields of technology, from domestic applications to the scientific and military solutions. Currently there are four most popular methods of 3D printing which can be distinguished: selective laser sintering, inkjet printing, Fused Deposition Modeling (FDM) and stereolithography [9]. One of the cheapest and most broadly widespread methods is the FDM, mainly due to the relatively easy realization of the printer on your own. Nevertheless, some other approaches are developed independently e.g. "OLO - The First Ever Smartphone 3D Printer" project published and development on the "Kickstarter" website. This handy 3D printer uses light display of the mobile phone which hardens special photopolymer resins.

There are numerous versions of available printers utilizing the FDM method both on the GPL licenses as well as based on the Creative Commons ones (e.g. CC-BY-NC-SA). Unfortunately, such devices often have more or less significant problems with the quality of printed elements. Technically, they are caused by poor quality materials used for the construction of the printing devices as well as some mistakes made during their development.

© Springer International Publishing AG 2017
R.S. Choraś (ed.), *Image Processing and Communications Challenges 8*,
Advances in Intelligent Systems and Computing 525, DOI 10.1007/978-3-319-47274-4_12

Another source of problems with the quality of 3D prints is the quality of the filament as well as the presence of some distortions during the printing process. Such types of contaminations can be intentionally introduced during printing in order to obtain lower quality 3D prints, further used for comparisons or the development of automatic quality assessment methods of 3D prints. As this issue can be considered as one of the most up-to-date challenges for image analysis, an approach based on the modern image quality metrics leading to very promising experimental results, is presented in this paper.

The application of machine vision system for monitoring the overall process of 3D printing has been presented in the article [2] whereas the observation of the top surface of 3D print has been discussed in the paper [3]. A comparative study of such approaches for fault detection can be found in some of the recent papers [1,8] as well as an initial study related to application of image analysis algorithms presented by Straub [7]. Nevertheless, such systems do not utilize any methods related to image quality assessment which can be useful for evaluation of the quality of the printed surface of a 3D object as discussed in our paper.

## 2   Similarity Based Image Quality Metrics

Starting from the idea of Universal Image Quality Index proposed in 2002 [10] and its further extension into Structural Similarity [11] and its multi-scale version [12], many image quality metrics have been proposed during recent years. Some of them are based on the general idea od the SSIM metric with some changes related to the use of e.g. gradients, variances, wavelet coefficients etc.

An interesting idea, partially based on the same scheme of calculations, has been proposed by Zhang [14] suggesting the use of Riesz transform. The authors have assumed a great importance of edge surrounding regions for the perceived image quality as well as some other key locations such as corners or zero-crossings. In order to detect those feature the use of Riesz transform has been proposed and therefore the first and second order Riesz transform coefficients have been used for the comparison of feature maps of two images (the reference image and distorted one). The local similarity index between two images, denoted as $f$ and $g$ respectively, can be expressed as:

$$d_i(x, y) = \frac{2 \cdot f_i(x, y) \cdot g_i(x, y) + C}{f_i^2(x, y) + g_i^2(x, y) + C} \tag{1}$$

where $i = 1..5$ are the numbers of five Riesz transform features and $C$ is a small constant value preventing the possible division by zero in the same way as initially proposed for the SSIM index. The overall RFSIM index is defined according to the following formula:

$$\text{RFSIM} = \prod_{i=1}^{5} \frac{\sum_x \sum_y d_i(x,y) \cdot M(x,y)}{\sum_x \sum_y M(x,y)} \tag{2}$$

where $M$ is the binary mask being the effect of edge detection.

Another metric proposed in 2011 by the same authors [13], namely Feature Similarity (FSIM), can be considered as one of the best full-reference image quality metrics according to the correlation with subjective perception of various kinds of image distortions. It utilizes the visual information being the combination of gradient magnitude and phase congruency [6] for calculation of the local similarity index using the formula:

$$S(x,y) = \left( \frac{2 \cdot PC_1(x,y) \cdot PC_2(x,y) + T_{PC}}{PC_1^2(x,y) + PC_2^2(x,y) + T_{PC}} \right) \cdot \left( \frac{2 \cdot G_1(x,y) \cdot G_2(x,y) + T_G}{G_1^2(x,y) + G_2^2(x,y) + T_G} \right) \tag{3}$$

where $T_{PC}$ and $T_G$ are small values playing the same role as their equivalents in SSIM and RFSIM metrics. For simplicity an equal importance of both factors (phase congruency $PC$ and gradient $G$) is assumed. For gradient computations different convolution filters may be applied (e.g. Sobel or Prewitt) but Scharr filter is recommended by inventors of the FSIM metric.

The overall quality index is obtained by averaging as:

$$\text{FSIM} = \frac{\sum_x \sum_y S(x,y) \cdot PC_m(x,y)}{\sum_x \sum_y PC_m(x,y)} \tag{4}$$

assuming $PC_m(x,y) = max(PC_1(x,y), PC_2(x,y))$ being the higher of the two local phase congruency values calculated for the reference and distorted images.

## 3    Proposed Approach and Discussion of Experiments

Usefulness of RFSIM and FSIM metrics discussed above for the quality evaluation of 3D prints can be considered in view of their potential application as no-reference ("blind") metrics. Since their typical usage is based on the comparison of two images (full-reference approach), it should be noticed that for the evaluation of 3D prints the reference 3D print can be often unavailable. Therefore a no-reference metric which does not require the knowledge of a perfect quality reference image would be much more desired.

Nevertheless, since both metrics produce local similarity indexes which are then averaged or combined in order to obtain the overall quality index, their calculation for some parts of the 3D prints is also possible. In the proposed approach a division of the image representing side view of the 3D printed surface used for testing purposes into several parts is assumed and then the FSIM and RFSIM metrics can be calculated between those fragments.

**Fig. 1.** Exemplary high (left column) and low quality (middle and right column) 3D prints used in experiments

Assuming the division into four parts of the same resolution, 6 such indexes can be obtained (calculated between the parts: 1 and 2, 1 and 3, 1 and 4, 2 and 3, 2 and 4, as well as 3 and 4 respectively). Those six values can be averaged leading to the overall no-reference index. For the image representing a perfect quality 3D print those values should be close to 1 due to high self-similarity of the image. In the presence of local visual distortions the values of quality metrics should decrease and therefore the lower quality prints could be identified.

The experimental verification of the proposed approach has been done by calculations of the metrics' values for several 3D prints of flat plates obtained using different filaments (some of them made from partially transmissive materials). For each color of the filament both high and low quality prints have been produced using two FDM printers (RepRap Pro Ormerod 2 and Prusa i3). Forcing the decrease of the 3D prints has been made mainly by modifications of the speed of the filament's delivery as well as changing the temperature. Some exemplary high and low quality 3D prints obtained by a digital camera and using a flatbed scanner are shown in Fig. 1. As can be easily noticed some differences in color may appear between the images captured by a camera and the scanned ones.

As both investigated metrics have been proposed mainly for quality evaluation of grayscale images, the impact of the color to grayscale conversion method on the obtained results has been investigated as well. During the calculations of the metrics' values some typical conversion methods based on the most popular color models have been used, including two ITU recommendations: BT.601-7 [5] for SDTV and BT.709-5 [4] for HDTV as well as the luminance components from the CIELAB and HSV color models.

## 4    Experimental Results

Unfortunately the application of RFSIM metric does not lead to satisfactory results as the values of this metric differ significantly both for high and low quality 3D prints. However, the use of Feature Similarity leads to much more promising results presented in detailed way in Table 1.

Analyzing the results obtained for the FSIM metric some interesting conclusions can be drawn. Classification of the 3D prints into high and low quality ones based on the FSIM results can be efficiently made for green, silver, white and red filaments. Nevertheless, a proper choice of threshold and the color to grayscale conversion may be different for each color of the filament.

For green 3D prints the best results can be obtained using the HSV and ITU BT.709-5 color models whereas the choice of ITU BT.601-7 may lead to some troubles in proper classification as the difference between the FSIM values for high and low quality prints is much smaller. The use of CIELAB color space leads to even worse results and improper classification.

For the silver filament only the CIELAB color space leads to useless results. The use of remaining three conversion methods with the threshold of the FSIM about 0.7 (similarly as for the two best conversions for green prints) allows the proper detection of the 3D prints quality.

A similar situation takes place for white prints as only CIELAB color space cannot be effectively used. Nevertheless, the threshold values should be increased to about 0.8.

The use of the red filament requires the use of the HSV color model but the choice of a proper threshold value at about 0.8 does not guarantee proper classification results.

The most troublesome situation took place for the orange filament as none of the color to grayscale conversion methods allowed the proper recognition between the high and low quality prints. Such a situation may be caused by different properties of this material as the light transmission coefficient for this material is much higher. This issue has been confirmed experimentally by measurements of the light intensity passing through the 3D printed plates of the same thickness and different colors. An illustration of this phenomenon is shown in Fig. 2.

**Table 1.** FSIM values obtained during experiments assuming the division of the image into 4 parts

| Color | Quality | FSIM values | | | |
|---|---|---|---|---|---|
| | | ITU BT.601-7 | ITU BT.709-5 | HSV | CIELAB |
| Green | High | 0.7843 | 0.7976 | 0.8287 | 0.9510 |
| | High | 0.7978 | 0.8083 | 0.8329 | 0.9853 |
| | High | 0.6857 | 0.7068 | 0.7507 | 0.7075 |
| | High | 0.8353 | 0.8487 | 0.8733 | 0.9845 |
| | High | 0.8164 | 0.8304 | 0.8581 | 0.9896 |
| | Low | 0.5872 | 0.5847 | 0.5809 | 0.7689 |
| | Low | 0.5728 | 0.5698 | 0.5664 | 0.7426 |
| | Low | 0.6608 | 0.6551 | 0.6424 | 0.8466 |
| | Low | 0.6605 | 0.6552 | 0.6416 | 0.7517 |
| Silver | High | 0.7245 | 0.7241 | 0.7253 | 0.6078 |
| | High | 0.7660 | 0.7659 | 0.7676 | 0.5251 |
| | High | 0.7703 | 0.7701 | 0.7702 | 0.6512 |
| | High | 0.7792 | 0.7784 | 0.8046 | 0.7429 |
| | High | 0.7519 | 0.7514 | 0.7726 | 0.7102 |
| | Low | 0.6237 | 0.6234 | 0.6267 | 0.4593 |
| | Low | 0.6723 | 0.6724 | 0.6753 | 0.5119 |
| | Low | 0.6834 | 0.6832 | 0.6805 | 0.6018 |
| | Low | 0.6789 | 0.6787 | 0.6744 | 0.6185 |
| Red | High | 0.8762 | 0.8572 | 0.9400 | 0.9923 |
| | High | 0.8131 | 0.7950 | 0.9021 | 0.9915 |
| | High | 0.6863 | 0.6789 | 0.7805 | 0.8551 |
| | High | 0.7029 | 0.6965 | 0.8103 | 0.7709 |
| | Low | 0.6664 | 0.6526 | 0.7211 | 0.5829 |
| | Low | 0.6737 | 0.6640 | 0.7545 | 0.7148 |
| | Low | 0.7048 | 0.6946 | 0.7562 | 0.9720 |
| | Low | 0.7691 | 0.7563 | 0.7932 | 0.9677 |
| White | High | 0.8794 | 0.8787 | 0.8792 | 0.3924 |
| | High | 0.8816 | 0.8824 | 0.8817 | 0.3929 |
| | High | 0.9435 | 0.9301 | 0.9577 | 1.0000 |
| | High | 0.9528 | 0.9358 | 0.9610 | 1.0000 |
| | Low | 0.7867 | 0.7871 | 0.7876 | 0.3957 |
| | Low | 0.7829 | 0.7831 | 0.7832 | 0.4524 |
| | Low | 0.7169 | 0.7115 | 0.7701 | 0.9980 |
| | Low | 0.6299 | 0.6186 | 0.6721 | 0.9999 |
| Orange | High | 0.7184 | 0.7158 | 0.7216 | 0.4125 |
| | High | 0.5880 | 0.5854 | 0.6669 | 0.4925 |
| | High | 0.7265 | 0.7212 | 0.7632 | 0.8898 |
| | High | 0.8237 | 0.8245 | 0.8347 | 0.9701 |
| | Low | 0.7798 | 0.7787 | 0.8021 | 0.7772 |
| | Low | 0.6771 | 0.6744 | 0.7334 | 0.6686 |
| | Low | 0.7215 | 0.7192 | 0.7463 | 0.8440 |

**Fig. 2.** Illustration of semi-transparency of different types of PLA filaments illuminated from back side

## 5   Concluding Remarks and Directions of Future Work

Quality assessment of 3D prints is an interesting field of further research. Starting from very promising results presented in this paper, further development of automatic quality estimation of 3D prints may be related to the use of some hybrid approaches based on combination of several metrics used for general purpose image quality assessment. Obtained results may be also used for on-line quality monitoring systems for 3D printers equipped with a camera with possibilities of corrections of some detected imperfections during the printing process.

Another possible direction of future work can be the increase of the versatility of the proposed approach in order to evaluate the quality of 3D prints of various shapes. This issue may be especially demanding due to potential changes of lighting conditions for different parts of the 3D objects visible on the images captured by a camera.

In order to obtain even better results a combination of color information from different RGB channels or some other color spaces may be also used. Assuming the availability of an additional light source mounted on the opposite side of the print relatively to the camera, such an approach may be combined with estimation of the semi-transparency level of the filament as well.

## References

1. Chauhan, V., Surgenor, B.: A comparative study of machine vision based methods for fault detection in an automated assembly machine. Procedia Manufact. **1**, 416–428 (2015)
2. Cheng, Y., Jafari, M.A.: Vision-based online process control in manufacturing applications. IEEE Trans. Autom. Sci. Eng. **5**(1), 140–153 (2008)
3. Fang, T., Jafari, M.A., Bakhadyrov, I., Safari, A., Danforth, S., Langrana, N.: Online defect detection in layered manufacturing using process signature. In: Proceedings of IEEE International Conference on Systems, Man and Cybernetics, San Diego, California, USA, vol. 5, pp. 4373–4378, October 1998

4. International Telecommunication Union: Recommendation BT.709-5 - Parameter values for the HDTV standards for production and international programme exchange (2002)
5. International Telecommunication Union: Recommendation BT.601-7 - Studio encoding parameters of digital television for standard 4:3 and wide-screen 16:9 aspect ratios (2011)
6. Liu, Z., Laganière, R.: Phase congruence measurement for image similarity assessment. Pattern Recogn. Lett. **28**(1), 166–172 (2007)
7. Straub, J.: Initial work on the characterization of additive manufacturing (3D printing) using software image analysis. Machines **3**(2), 55–71 (2015)
8. Szkilnyk, G., Hughes, K., Surgenor, B.: Vision based fault detection of automated assembly equipment. In: Proceedings of ASME/IEEE International Conference on Mechatronic and Embedded Systems and Applications, Parts A and B, Washington, DC, USA, vol. 3, pp. 691–697, August 2011
9. Tourloukis, G., Stoyanov, S., Tilford, T., Bailey, C.: Data driven approach to quality assessment of 3D printed electronic products. In: Proceedings of 38th International Spring Seminar on Electronics Technology (ISSE), Eger, Hungary, pp. 300–305, May 2015
10. Wang, Z., Bovik, A.: A universal image quality index. IEEE Sig. Process. Lett. **9**(3), 81–84 (2002)
11. Wang, Z., Bovik, A.C., Sheikh, H., Simoncelli, E.: Image quality assessment: from error measurement to structural similarity. IEEE Trans. Image Process. **13**(4), 600–612 (2004)
12. Wang, Z., Simoncelli, E., Bovik, A.C.: Multi-scale structural similarity for image quality assessment. In: Proceedings of the 37th IEEE Asilomar Conference on Signals, Systems and Computers, Pacific Grove, California (2003)
13. Zhang, L., Zhang, L., Mou, X., Zhang, D.: FSIM: a feature similarity index for image quality assessment. IEEE Trans. Image Process. **20**(8), 2378–2386 (2011)
14. Zhang, L., Zhang, L., Mou, X.: RFSIM: A feature based image quality assessment metric using Riesz transforms. In: Proceedings of the 17th IEEE International Conference on Image Processing, Hong Kong, China, pp. 321–324 (2010)

# Noise Objects Tracking Using Multiple Order Statistics and Spatio–Temporal Track–Before–Detect Algorithm

Przemysław Mazurek[✉]

Department of Signal Processing and Multimedia Engineering,
West–Pomeranian University of Technology,
26. Kwietnia 10 St., 71126 Szczecin, Poland
przemyslaw.mazurek@zut.edu.pl

**Abstract.** Low SNR object could be tracked and detected using Track–Before–Detect (TBD) algorithms. Most TBD algorithm assume positive signal of object and Gaussian background noise. Multiple order statistics (mean, variance and skewness) for the improving of detection are proposed and analyzed in this paper. Monte Carlo results and the dependence between mean, standard deviation and skewness are provided.

**Keywords:** Tracking · Track–before–detect · Skewness

## 1 Introduction

Tracking systems are applied in numerous applications [1]. They are important part of the air, water, underwater and space surveillance systems. The object tracking is quite simple if the object signal is significantly distinguished from the background. The high signal to noise ratio ($SNR$) case allows the application of fixed or adaptive threshold algorithms for the estimation of object's position. Reduced SNR cases require [1] the application of predictors, like the Benedict–Bordner, the Kalman [1] or the Bayes [1,13] filters. The tracking filter reduces the influence of false observations. The restoration of the object position, even if the correct observation is not detected in the current measurement is possible, by the prediction. Typical tracking system uses the detection and tracking approach.

Multiple object tracking systems requires an additional assignment algorithm. The assignment algorithm is responsible for the track maintenance tasks: initiation, removal, merging and assignment of the observation to the appropriate predictor [1]. It improves tracking quality of overall system, also.

The conventional detection and tracking approach [1,13] is not adequate for very low $SNR$ cases ($SNR < 1$). The amount of false observations, obtained by the threshold algorithm is huge, so the selection of observed object for the predictor is not possible.

© Springer International Publishing AG 2017
R.S. Choraś (ed.), *Image Processing and Communications Challenges 8*,
Advances in Intelligent Systems and Computing 525, DOI 10.1007/978-3-319-47274-4_13

Alternative Track–Before–Detect (TBD) approach allows the tracking of objects that are below the noise floor ($SNR < 1$) [1,2,13]. The first operation is based on the tracking with the use of raw measurements [13]. All possible trajectories are processed, even if no object is in the range. Multidimensional spatial and temporal filtering in TBD algorithm allows the noise reduction and improves $SNR$ value for the true object. The second operation is the detection using obtained multidimensional state–space for the estimation of position and motion together.

### 1.1   Outline of the Paper

The Spatio–Temporal Track–Before–Detect algorithm is applied for the positive signals typically, but noise objects require additional preprocessing. The proposed processing technique for noise objects is discussed in Sect. 2. Multiple order statistics are proposed in Sect. 3. Results of Monte Carlo tests are provided in Sect. 4 and discussion is in Sect. 5. Conclusions and further works are considered in Sect. 6.

### 1.2   Related Works

Segmentation based on high order statistic is considered in [11]. Local standard deviation, absolute value of mean with dot product preprocessing [7], maximal autocovariance [10], chi–square statistic are proposed in the previous works. Filter banks are considered in [9]. Two dimensional TBD tracking allows segmentation [4,12] in medical applications also. High order statistics are important in SAR image segmentation [14].

### 1.3   Contribution of the Paper

Main contribution of the paper is the proposal multiple order statistics for the preprocessing of measurement. Previous works related to this topic area are related with the application of the variance and distributions comparison only. Spatio–Temporal Track–Before–Detect (ST TBD) algorithm is considered in [6,8] for example.

## 2   Noise Objects

Typical object signature is the positive signal (reflected or emitted light in visible light or infrared vision systems) that could be processed by ST TBD directly. Numerous stealth technologies are applied for the reduction of positive and modulated signals.

Noise objects are specific class of signals, where the object's signal is noise only, disturbed by the background noise. Only noises are measured and the differences between them could be very small, so object noise is hidden in the

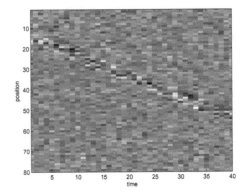

**Fig. 1.** Example of noise signal $s = 1$ disturbed by additive noise $s = 1$

background noise. The TBD algorithm is necessary for such scenarios together with the proper preprocessing of measurements.

In this paper the spatial noise is assumed, so the object is the extended target [1] that occupies a few measurement cells (pixels). Example 1D measurements (for simplification) are shown in Fig. 1. The background clutter is additive Gaussian noise ($s = 1$) and the object is Gaussian noise (7 pixels width), but different distributions are possible to process.

## 2.1 Detection of Noise Objects

The estimation of statistical properties in the spatial domain using moving window approach could be applied. The size of the window should be similar to the size of object. Too large size reduces spatial resolution of the system and too small size reduces sensitivity. The parametric techniques estimate known a priori distribution parameters, related to the background or object (e.g. mean, variance). The non–parametric techniques allows the comparison between two distributions (global related to the overall measurements and local related to the moving window position). The prepossessing of measurements is necessary for spatial difference estimation (Fig. 2). The ST TBD algorithm allows the detection of positive signal, so zero mean background and object noises cannot be distinguished without parametric or non–parametric estimation.

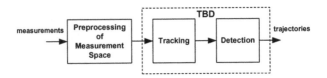

**Fig. 2.** Preprocessing and ST TBD algorithms

The moving window size $2L + 1$ is applied for measured signal $M$ using the following formula:

$$X(k, s) = std\,[M(k, s - L), \cdots, M(k, s + L)] \tag{1}$$

## 3   Multiple Order Statistics

The mean $\bar{x}$ - 1'st raw moment (2), and sample variance $S^2$ - 2'nd central moment (3) are basic descriptors of distribution [5]. The sample skewness [3] $G_1$ - 3'rd standardized moment (5) could be applied for the improving of distribution description.

$$\bar{x} = \frac{1}{n} \sum_{i=1}^{n} x_i \tag{2}$$

$$S^2 = \frac{1}{n-1} \sum_{i=1}^{n} (x_i - \bar{x})^2 \tag{3}$$

$$g_1 = \frac{\frac{1}{n} \sum_{i=1}^{n} (x_i - \bar{x})^3}{(\frac{1}{n} \sum_{i=1}^{n} (x_i - \bar{x})^2)^{3/2}} \tag{4}$$

$$G_1 = \frac{\sqrt{n(n-1)}}{n-2} g_1 \tag{5}$$

The properties moments applied separately are evaluated using Monte Carlo tests. The minimum sample size for using a parametric statistical test varies among texts. Smaller sample sizes (typically $< 30$) give very weak descriptors of distribution. The ST TBD work as filter that denoises descriptors intentionally and improves descriptors by spatio–temporal processing.

## 4   Monte Carlo Tests

The 1D tracking scenario is assumed and there are 1200 measurement cells. The extended object occupies 7 cells and this width is known a priori for the preprocessing algorithm. The background is filled by the Gaussian noise. Object signal replace cells of the background, so is not additive. The parameters of noise object are tested for specified ranges. Mean error of the position error for multiple tests is computed. The velocity of object is from 0 to 10 range with integer values. There are no transitions between velocity subspaces, because the velocity is fixed. Two smoothing coefficient values are tested: $\alpha = 0.95$ and $\alpha = 0.98$.

The results show not only changes for a single variable parameter, but for all tested moments. There are three types of errors: maximal, mean and median (Fig. 5).

## 5   Discussion

There are not necessary additional assumptions related to the shapes of distribution for non–parametric methods using the proposed statistics (Fig. 8).

The obtained results show the correctness of proposed estimators and limitations of the method. Zero distance error values could be obtained for all estimators. The most interesting result is the dependence between mean, variance
and skewness. This is the result of boundary position of window of analysis. It
is visible for variable mean especially in Figs. 3 and 6. All considered estimators
change value for such case. There are similar results for variable standard deviation and mean estimator (Figs. 4 and 7), but this influence is very low. The
application of the higher $\alpha$ value improves the sensitivity of considered method,
that is expected results. Small number of samples for the estimation of parameters gives error that is reduced but TBD algorithm due to recursive processing.

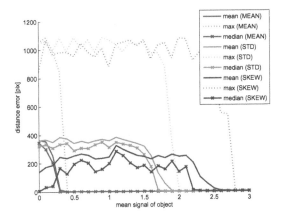

**Fig. 3.** Position errors for variable mean value and $\alpha = 0.95$

Data fusion could be applied for conversion of obtained estimates to the single
measurement space of ST TBD using weighted formula:

$$\begin{aligned}
X^2(k, s) = \; & w_1(\bar{x}(k, M(s - L, \cdots, s + L)) - \bar{x}(k, M))^2 \\
& + w_2(S^2(k, M(s - L, \cdots, s + L)) - S^2(k, M))^2 \\
& + w_3(G_1(k, M(s - L, \cdots, s + L)) - G_1(k, M))^2
\end{aligned} \tag{6}$$

but this topic is not addressed in this paper. The knowledge about object and
background noise characteristics could be incorporated by the optimization of
the weight values ($w_i$) for improving divergence between the background and
object.

## 6   Conclusions and Further Works

The detection and tracking of noise object could be applied for image processing
task, also. Mean and variance are visible for humans for high SNR scenarios, but

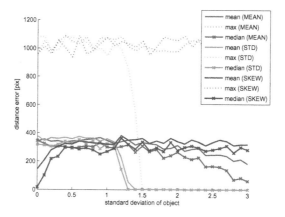

**Fig. 4.** Position errors for variable standard deviation value and $\alpha = 0.95$

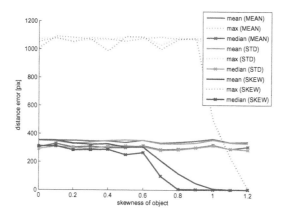

**Fig. 5.** Position errors for variable skewness value and $\alpha = 0.95$

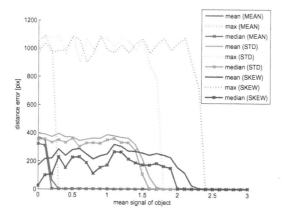

**Fig. 6.** Position errors for variable mean value and $\alpha = 0.98$

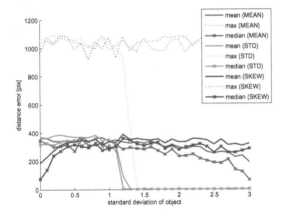

**Fig. 7.** Position errors for variable standard deviation value and $\alpha = 0.98$

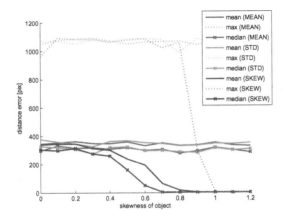

**Fig. 8.** Position errors for variable skewness value and $\alpha = 0.98$

higher order differences are not visible. The detection of lines is an interesting application.

ST TBD algorithm requires single measurement space, so weighted preprocessing is interesting, but the relation between value weights should be established. Alternative approach based on multiple ST TBD algorithms (every algorithm for particular statistical moment) with the data fusion of outputs is possible also, but requires more computations.

Obtained results show the important behavior of moving window. The estimation is correct if the window is over the objects measurement space cells. Additional changes occur if the window overlaps of objects and background cells together. This property should be considered by the data fusion algorithm.

**Acknowledgment.** This work is supported by the UE EFRR ZPORR project Z/2.32/I/1.3.1/267 /05 "Szczecin University of Technology – Research and Education Center of Modern Multimedia Technologies" (Poland).

# References

1. Blackman, S., Popoli, R.: Design and Analysis of Modern Tracking Systems. Artech House, Norwood (1999)
2. Boers, Y., Ehlers, F., Koch, W., Luginbuhl, T., Stone, L., Streit, R.: Track before detect algorithms. EURASIP J. Adv. Signal Process. **2008**, 2 (2008). doi:10.1155/2008/413932. Article ID 413932
3. Doane, D., Seward, L.: Measuring skewness: a forgotten statistic? J. Stat. Educ. **19**(2), 1–18 (2011)
4. Habrat, K., Habrat, M., Gronkowska-Serafin, J., Piórkowski, A.: Cell detection in corneal endothelial images using directional filters. In: Choraś, R.S. (ed.) Image Processing and Communications Challenges 7. AISC, vol. 389, pp. 113–123. Springer, Heidelberg (2016). doi:10.1007/978-3-319-23814-2_14
5. Kendall, M., Stuart, A.: The Advanced Theory of Statistics, Distribution Theory, vol. 1. Griffin, London (1969)
6. Mazurek, P.: Optimization of bayesian track-before-detect algorithms for GPGPUs implementations. Electr. Rev. R **86**(7), 187–189 (2010)
7. Mazurek, P.: Application of dot product for track-before-detect tracking of noise objects. Poznań Univ. Technol. Acad. J. Electr. Eng. **76**, 101–107 (2013)
8. Mazurek, P.: Code reordering using local random extraction and insertion (LREI) operator for GPGPU-based track-before-detect systems. Soft Comput. **18**(6), 1095–1106 (2013)
9. Mazurek, P.: Track-before-detect filter banks for noise object tracking. Int. J. Electron. Telecommun. **59**(4), 325–330 (2013)
10. Mazurek, P.: Preprocessing using maximal autocovariance for spatio-temporal track-before-detect algorithm. Adv. Intell. Syst. Comput. **233**, 45–54 (2014)
11. Petrou, M., Kovalev, V., Reichenbach, J.: High order statistics for tissue segmentation. In: Handbook of Medical Image Processing and Analysispp, pp. 245–257. Academic Press (2009)
12. Scott, T.A., Nilanjan, R.: Biomedical Image Analysis: Tracking. Morgan & Claypool, San Rafael (2005)
13. Stone, L., Barlow, C., Corwin, T.: Bayesian Multiple Target Tracking. Artech House, Norwood (1999)
14. Sui, H., Xu, C., Liu, J., Sun, K., Wen, C.: A novel multi-scale level set method for sar image segmentation based on a statistical model. Int. J. Remote Sens. **33**(17), 5600–5614 (2012)

# Active Learning Algorithm Using the Discrimination Function of the Base Classifiers

Robert Burduk[✉]

Department of Systems and Computer Networks, Wroclaw University of Technology,
Wybrzeze Wyspianskiego 27, 50-370 Wroclaw, Poland
robert.burduk@pwr.edu.pl

**Abstract.** The goal of the Active Learning algorithm is to reduce the number of labeled examples needed for learning. In this paper we propose the new AL algorithm based on the analysis of decision profiles. The decision profiles are obtained from the outputs of the base classifiers that form an ensemble of classifiers. The usefulness of the proposed algorithm is experimentally evaluated on several data sets.

**Keywords:** Active Learning · Query by Committee · Multiple classifier system

## 1 Introduction

Supervised learning is one of the types of machine learning [1]. Classification methods are applied in many practical tasks [5, 8, 10]. Generally, the recognition algorithm maps the feature space to the set of class labels. This process requires a sufficiently large number of training examples in the learning set. Typically these examples are manually labelled by the expert (sometimes called - oracle). On the other hand unlabelled examples are much easier to be acquired.

Active Learning (AL) [6] is a special case of machine learning in which a learning algorithm is able to interactively query the expert to obtain a label for unlabelled examples. These labelled examples are further used to improve a classifier. The key issue is to select the most informative examples. In this paper we use Query by Committee (QBC) approach [11]. This approach to AL is based on using the ensemble of classifiers to select the most informative examples.

The text is organized as follows: after this introduction, in Sect. 2 the idea of an ensemble of classifiers is presented. Section 3 contains the description of the proposed AL scheme based on the QBC approach. The experimental results on several data sets are presented in Sect. 4. Finally, conclusions from the experiments and future research proposals are presented.

## 2 Ensemble of Classifiers

The classification task can be accomplished by a single classifier or by a team of classifiers. In the literature, the use of the multiple classifiers for a decision

© Springer International Publishing AG 2017
R.S. Choraś (ed.), *Image Processing and Communications Challenges 8*,
Advances in Intelligent Systems and Computing 525, DOI 10.1007/978-3-319-47274-4_14

problem is known as the multiple classifier systems (MCS) or the ensemble of classifiers EoC [7,12]. The construction of MSC consists of three phases: generation, selection and integration [3]. In the second phase, which is discussed in this paper, one classifier or a subset of the base classifiers is selected to make the final decision which is to assign an object to the class label.

The output of an individual classifier can be divided into three types [13]:

- The abstract level – the classifier $\psi$ assigns the unique label $j$ to the given input $x$.
- The rank level – in this case for each input $x$, each classifier produces an integer rank array. Each element within this array corresponds to one of the defined class labels. The array is usually sorted with the label at the top being the first choice.
- The measurement level – the output of a classifier is represented by a measurement value that addresses the degree of assigning the class label to the given output $x$. An example of such a representation of the output is a posteriori probability returned by Bayes classifier.

Let us assume that we possess $K$ of different classifiers $\Psi_1, \Psi_2, \ldots, \Psi_K$. Such a set of classifiers, which is constructed on the basis of the same learning sample is called an ensemble of classifiers or a combining classifier. However, any of $\Psi_i$ classifiers is described as a component or a base classifier. As a rule $K$ is assumed to be an odd number and each of $\Psi_i$ classifiers makes an independent decision. As a result, of all the classifiers' actions, their $K$ responses are obtained. Having at the disposal a set of base classifiers one should determine the procedure of making the ultimate decision regarding the allocation of the object to the given class. It implies that the output information from all $K$ component classifiers is applied to make the ultimate decision.

In this work we consider the situation when each base classifier returns the estimation of a posteriori probability. This means that the output of all the base classifiers is at the measurement level. Let us denote a posteriori probability estimation (most often discrimination function – DF) by $p_k(\omega|x)$, $k = 1, 2, \ldots, K$, $\omega = 1, 2, \ldots, \Omega$, where $\Omega$ is the number of the class labels. One of the possible methods for such outputs is the linear combination method. This method makes use of the linear function like Sum, Prod or Mean for the combination of the outputs. In the sum method the score of the group of classifiers is based on the application of the following sums:

$$s_i(x) = \sum_{k=1}^{K} p_k(\omega|x), \qquad \omega = 1, 2, \ldots, \Omega. \tag{1}$$

The final decision of the group of classifiers is made following the maximum rule and is presented accordingly, depending on the sum method (1):

$$\Psi_S(x) = \arg\max_i s_i(x). \tag{2}$$

In the presented method (2) DF obtained from the base classifiers take an equal part in building MCSs. This is the simplest situation in which we do not need additional information on the testing process of the base classifiers except for the models of these classifiers. One of the possible methods in which weights of the base classifier are used is presented in [4].

## 3    Proposal of Active Learning Algorithm Using the Decision Profile

The general AL scheme is defined as follows [15]:

**Input:** Learning algorithm - $A$; Set of labeled training examples - $L$; Set of unlabeled training examples - $U$; Number of active learning iterations - $n$; Number of selected examples - $m$.

**Repeat** $n$ times

1. Generate a committee of classifiers, $C^* = EnsembleMethod(A, L)$
2. $\forall x_j \in U$ compute $InformationValue(C^*, x_j)$, based on the current committee
3. Select a subset $S$ of $m$ examples that are the most informative
4. Obtain a label for examples in $S$ from "oracle" or an expert
5. Remove examples in $S$ from $U$ and add to $L$

**Return** $EnsembleMethod(A, L)$

### 3.1    Proposal of Information Value Calculations

For $K$ base classifier their outputs are arranged in the decision profile:

$$DP(x) = \begin{bmatrix} p_1(1|x) & \vdots & p_1(\Omega|x) \\ \vdots & \vdots & \vdots \\ p_K(1|x) & \vdots & p_K(\Omega|x) \end{bmatrix}. \tag{3}$$

During learning of the base classifiers we obtain $m$ decision profiles, where $m$ is the number of objects from the learning set. From the decision profiles we calculate the decision scheme according to the formula:

$$DS = \begin{bmatrix} ds_{11} & \vdots & ds_{1\Omega} \\ \vdots & \vdots & \vdots \\ ds_{K1} & \vdots & ds_{K\Omega} \end{bmatrix}, \tag{4}$$

where

$$ds_{k\omega} = \frac{\sum_{n=1}^{m} I(\Psi_k(x_n) = \omega_n)\, p_k(\omega_n|x_n)}{\sum_{n=1}^{m} I(\Psi_k(x_n) = \omega_n)}, \tag{5}$$

where $I(\cdot)$ is the indicator function. The value of $ds_{k\omega}$ is calculated only from those DFs for which the classifier $k$ did not make an error.

The information value is calculated for the new object on the basis of its decision profile according to the formula:

$$IV(x) = \sum_{k=1}^{K} \sum_{\omega=1}^{\Omega} I(p_k(\omega|x) < ds_{k\omega}). \tag{6}$$

The obtained information value $IV$ for the new object from a set of an unlabelled training example is used to indicate the object $x$ to be labelled. The labelled object is the one that exceeds the established value of the $IV$. In the experiments the algorithm using the proposed above method is denoted as $\Psi_{AL}^{IV}$.

## 4   Experimental Studies

In the experiential research we used two data sets from the UCI repository [9], one data set form Keel repository and the two generated randomly – they are the so called Banana and Higleyman sets. The numbers of attributes and examples are presented in Table 1. In the experiments we have used the standard 10-fold-cross-validation method and the feature selection process [14] was not performed.

Table 1. Description of data sets selected for the experiments

| Data set | Example | Attribute |
|---|---|---|
| Banana | 2000 | 2 |
| Highleyman | 400 | 2 |
| MAGIC Gamma Telescope | 19020 | 11 |
| Phone | 961 | 6 |
| Pima Indians Diabetes | 768 | 8 |

In the experiments 9 base classifiers were used. One of them (labelled as $\Psi_1$) used the decision trees algorithms, with the splitting rule based on entropy, the number of branches equal to 2 and the depth of the precision tree having at most 6 levels. Two of them work according to $k-NN$ rule where $k$ parameter is equal to 3 or 5 and they are labelled as $\Psi_2$ and $\Psi_3$ respectively. The classifier labelled as $\Psi_4$ is the rule induction classifier. This classifier uses a tree-based modelling in which a tree is run and models for which the purity is at or above a specified threshold are removed from the training data set and placed on the side. The fifth classifier $\Psi_5$ uses Support Vector Machines models with the Decomposed Quadratic Programming estimation method. The sixth classifier $\Psi_6$ uses the least squares regression model to predict the class label. The last of the base classifiers $\Psi_7$ is the multilayer perceptron model with 3 hidden units.

Tables 2, 3, 4, 5, 6, 7, 8, 9, 10 and 11 show the results of the classification for all base classifiers and two ensemble methods. The results refer to the first iteration of AL scheme and are for information value equal 12 or 15.

**Table 2.** Classification accuracy for the base classifiers ($\Psi_1, ..., \Psi_9$), Majority Voting and Sum method for AL scheme with $IV = 12$ and random labelling of unlabelled training examples - Pima data set

| AL scheme | $\Psi_1$ | $\Psi_2$ | $\Psi_3$ | $\Psi_4$ | $\Psi_5$ | $\Psi_6$ | $\Psi_7$ | $\Psi_8$ | $\Psi_9$ | $\Psi_{MV}$ | $\Psi_{AL}^{12}$ |
|---|---|---|---|---|---|---|---|---|---|---|---|
| k = 0 | 0.25 | 0.25 | 0.22 | 0.21 | 0.40 | 0.28 | 0.26 | 0.27 | 0.28 | 0.18 | 0.21 |
| k = 1 | 0.22 | 0.22 | 0.40 | 0.27 | 0.22 | 0.27 | 0.25 | 0.31 | 0.30 | 0.26 | 0.25 |
| Random | 0.25 | 0.25 | 0.40 | 0.22 | 0.35 | 0.28 | 0.26 | 0.35 | 0.30 | 0.25 | 0.28 |

**Table 3.** Classification accuracy for the base classifiers ($\Psi_1, ..., \Psi_9$), Majority Voting and Sum method for AL scheme with $IV = 15$ and random labelling of unlabelled training examples - Pima data set

| AL scheme | $\Psi_1$ | $\Psi_2$ | $\Psi_3$ | $\Psi_4$ | $\Psi_5$ | $\Psi_6$ | $\Psi_7$ | $\Psi_8$ | $\Psi_9$ | $\Psi_{MV}$ | $\Psi_{AL}^{15}$ |
|---|---|---|---|---|---|---|---|---|---|---|---|
| k = 0 | 0.25 | 0.25 | 0.22 | 0.21 | 0.40 | 0.28 | 0.26 | 0.27 | 0.28 | 0.18 | 0.21 |
| k = 1 | 0.25 | 0.25 | 0.40 | 0.21 | 0.40 | 0.28 | 0.26 | 0.27 | 0.28 | 0.18 | 0.21 |
| Random | 0.34 | 0.34 | 0.40 | 0.23 | 0.35 | 0.43 | 0.26 | 0.25 | 0.23 | 0.30 | 0.26 |

**Table 4.** Classification accuracy for the base classifiers ($\Psi_1, ..., \Psi_9$), Majority Voting and Sum method for AL scheme with $IV = 12$ and random labelling of unlabelled training examples - Banana data set

| AL scheme | $\Psi_1$ | $\Psi_2$ | $\Psi_3$ | $\Psi_4$ | $\Psi_5$ | $\Psi_6$ | $\Psi_7$ | $\Psi_8$ | $\Psi_9$ | $\Psi_{MV}$ | $\Psi_{AL}^{12}$ |
|---|---|---|---|---|---|---|---|---|---|---|---|
| k = 0 | 0.28 | 0.28 | 0.45 | 0.42 | 0.16 | 0.14 | 0.27 | 0.14 | 0.13 | 0.21 | 0.12 |
| k = 1 | 0.18 | 0.18 | 0.45 | 0.51 | 0.12 | 0.14 | 0.32 | 0.12 | 0.13 | 0.15 | 0.13 |
| Random | 0.24 | 0.24 | 0.45 | 0.42 | 0.13 | 0.14 | 0.28 | 0.11 | 0.11 | 0.16 | 0.12 |

**Table 5.** Classification accuracy for the base classifiers ($\Psi_1, ..., \Psi_9$), Majority Voting and Sum method for AL scheme with $IV = 15$ and random labelling of unlabelled training examples - Banana data set

| AL scheme | $\Psi_1$ | $\Psi_2$ | $\Psi_3$ | $\Psi_4$ | $\Psi_5$ | $\Psi_6$ | $\Psi_7$ | $\Psi_8$ | $\Psi_9$ | $\Psi_{MV}$ | $\Psi_{AL}^{15}$ |
|---|---|---|---|---|---|---|---|---|---|---|---|
| k = 0 | 0.28 | 0.28 | 0.45 | 0.42 | 0.16 | 0.14 | 0.27 | 0.14 | 0.13 | 0.21 | 0.12 |
| k = 1 | 0.18 | 0.18 | 0.45 | 0.45 | 0.15 | 0.14 | 0.30 | 0.14 | 0.12 | 0.18 | 0.13 |
| Random | 0.31 | 0.31 | 0.45 | 0.41 | 0.17 | 0.14 | 0.25 | 0.13 | 0.13 | 0.22 | 0.12 |

**Table 6.** Classification accuracy for the base classifiers ($\Psi_1, ..., \Psi_9$), Majority Voting and Sum method for AL scheme with $IV = 12$ and random labelling of unlabelled training examples - Higleman data set

| AL scheme | $\Psi_1$ | $\Psi_2$ | $\Psi_3$ | $\Psi_4$ | $\Psi_5$ | $\Psi_6$ | $\Psi_7$ | $\Psi_8$ | $\Psi_9$ | $\Psi_{MV}$ | $\Psi_{AL}^{12}$ |
|---|---|---|---|---|---|---|---|---|---|---|---|
| k = 0 | 0.05 | 0.05 | 0.2 | 0.17 | 0.12 | 0.05 | 0.05 | 0.12 | 0.05 | 0.05 | 0.05 |
| k = 1 | 0.05 | 0.05 | 0.5 | 0.15 | 0.05 | 0.05 | 0.07 | 0.02 | 0.05 | 0.05 | 0.05 |
| Random | 0.07 | 0.07 | 0.5 | 0.2 | 0.12 | 0.07 | 0.1 | 0.05 | 0.05 | 0.05 | 0.05 |

**Table 7.** Classification accuracy for the base classifiers $(\Psi_1, ..., \Psi_9)$, Majority Voting and Sum method for AL scheme with $IV = 15$ and random labelling of unlabelled training examples - Higleman data set

| AL scheme | $\Psi_1$ | $\Psi_2$ | $\Psi_3$ | $\Psi_4$ | $\Psi_5$ | $\Psi_6$ | $\Psi_7$ | $\Psi_8$ | $\Psi_9$ | $\Psi_{MV}$ | $\Psi_{AL}^{15}$ |
|---|---|---|---|---|---|---|---|---|---|---|---|
| k = 0 | 0.05 | 0.05 | 0.2 | 0.17 | 0.12 | 0.05 | 0.05 | 0.12 | 0.05 | 0.05 | 0.05 |
| k = 1 | 0 | 0 | 0.5 | 0.17 | 0.12 | 0.1 | 0.12 | 0.07 | 0.1 | 0.1 | 0.1 |
| Random | 0.17 | 0.17 | 0.67 | 0.17 | 0.1 | 0.05 | 0.07 | 0.1 | 0.1 | 0.1 | 0.07 |

**Table 8.** Classification accuracy for the base classifiers $(\Psi_1, ..., \Psi_9)$, Majority Voting and Sum method for AL scheme with $IV = 12$ and random labelling of unlabelled training examples - Magic data set

| AL scheme | $\Psi_1$ | $\Psi_2$ | $\Psi_3$ | $\Psi_4$ | $\Psi_5$ | $\Psi_6$ | $\Psi_7$ | $\Psi_8$ | $\Psi_9$ | $\Psi_{MV}$ | $\Psi_{AL}^{12}$ |
|---|---|---|---|---|---|---|---|---|---|---|---|
| k = 0 | 0.16 | 0.16 | 0.20 | 0.20 | 0.20 | 0.18 | 0.17 | 0.20 | 0.20 | 0.16 | 0.16 |
| k = 1 | 0.15 | 0.15 | 0.36 | 0.20 | 0.21 | 0.17 | 0.18 | 0.19 | 0.19 | 0.16 | 0.15 |
| Random | 0.15 | 0.15 | 0.36 | 0.21 | 0.21 | 0.17 | 0.16 | 0.18 | 0.18 | 0.16 | 0.15 |

**Table 9.** Classification accuracy for the base classifiers $(\Psi_1, ..., \Psi_9)$, Majority Voting and Sum method for AL scheme with $IV = 15$ and random labelling of unlabelled training examples - Magic data set

| AL scheme | $\Psi_1$ | $\Psi_2$ | $\Psi_3$ | $\Psi_4$ | $\Psi_5$ | $\Psi_6$ | $\Psi_7$ | $\Psi_8$ | $\Psi_9$ | $\Psi_{MV}$ | $\Psi_{AL}^{15}$ |
|---|---|---|---|---|---|---|---|---|---|---|---|
| k = 0 | 0.16 | 0.16 | 0.20 | 0.20 | 0.20 | 0.18 | 0.17 | 0.20 | 0.20 | 0.16 | 0.16 |
| k = 1 | 0.15 | 0.15 | 0.36 | 0.25 | 0.19 | 0.18 | 0.16 | 0.20 | 0.19 | 0.15 | 0.14 |
| Random | 0.15 | 0.15 | 0.36 | 0.21 | 0.20 | 0.19 | 0.17 | 0.19 | 0.19 | 0.16 | 0.16 |

**Table 10.** Classification accuracy for the base classifiers $(\Psi_1, ..., \Psi_9)$, Majority Voting and Sum method for AL scheme with $IV = 12$ and random labelling of unlabelled training examples - Phone data set

| AL scheme | $\Psi_1$ | $\Psi_2$ | $\Psi_3$ | $\Psi_4$ | $\Psi_5$ | $\Psi_6$ | $\Psi_7$ | $\Psi_8$ | $\Psi_9$ | $\Psi_{MV}$ | $\Psi_{AL}^{12}$ |
|---|---|---|---|---|---|---|---|---|---|---|---|
| k = 0 | 0.25 | 0.25 | 0.26 | 0.27 | 0.30 | 0.26 | 0.21 | 0.26 | 0.26 | 0.23 | 0.22 |
| k = 1 | 0.22 | 0.22 | 0.32 | 0.27 | 0.17 | 0.19 | 0.20 | 0.13 | 0.14 | 0.17 | 0.15 |
| Random | 0.20 | 0.20 | 0.32 | 0.28 | 0.19 | 0.22 | 0.21 | 0.15 | 0.15 | 0.18 | 0.17 |

**Table 11.** Classification accuracy for the base classifiers $(\Psi_1, ..., \Psi_9)$, Majority Voting and Sum method for AL scheme with $IV = 15$ and random labelling of unlabelled training examples - Phone data set

| AL scheme | $\Psi_1$ | $\Psi_2$ | $\Psi_3$ | $\Psi_4$ | $\Psi_5$ | $\Psi_6$ | $\Psi_7$ | $\Psi_8$ | $\Psi_9$ | $\Psi_{MV}$ | $\Psi_{AL}^{15}$ |
|---|---|---|---|---|---|---|---|---|---|---|---|
| k = 0 | 0.25 | 0.25 | 0.26 | 0.27 | 0.30 | 0.26 | 0.21 | 0.26 | 0.26 | 0.23 | 0.22 |
| k = 1 | 0.23 | 0.23 | 0.32 | 0.27 | 0.23 | 0.20 | 0.22 | 0.2 | 0.21 | 0.21 | 0.19 |
| Random | 0.22 | 0.22 | 0.32 | 0.27 | 0.22 | 0.18 | 0.20 | 0.18 | 0.19 | 0.18 | 0.17 |

The obtained results are promising. For a given data sets we received improvement of classification quality. This improvement relates to the proposed AL algorithm compared to random labelling of the unlabelled training examples. For example, algorithm $\Psi_{AL}^{12}$ in our AL approach is about two percent higher than the same algorithm with random labelling in the case of MAGIC Gamma Telescope and Phone data sets. Made experiments also show that the selection of $IV$ parameter values significantly affect the quality of classification. For example algorithm $\Psi_{AL}^{IV}$ obtained the best results for $IV = 15$ and $IV = 12$ for MAGIC Gamma Telescope and Phone data sets respectively.

## 5   Conclusion

The paper presents the new AL algorithm based on the QBC approach. In the information value calculating process the decision profiles are used. In the paper experiments on several data sets were carried out. The obtained results are promising. This is due to the improvement of classification quality when we use the proposed AL algorithm compared to random labelling of unlabelled training examples.

In the future studies we plan to discuss the impact of using another value of the information value. In addition, further experiments should apply imbalanced data set [2] and not only to the binary data set.

**Acknowledgments.** This work was supported by the Polish National Science Center under the grant no. DEC-2013/09/B/ST6/02264.

## References

1. Bishop, C.M.: Pattern Recognition and Machine Learning. Information Science and Statistics. Springer, New York (2006)
2. Borowska, K., Topczewska, M.: New data level approach for imbalanced data classification improvement. In: Burduk, R., Jackowski, K., Kurzyński, M., Woźniak, M., Żołnierek, A. (eds.) Proceedings of the 9th International Conference on Computer Recognition Systems CORES 2015. Advances in Intelligent Systems and Computing, vol. 403, pp. 283–294. Springer, Switzerland (2016)
3. Britto, A.S., Sabourin, R., Oliveira, L.E.: Dynamic selection of classifiersa comprehensive review. Pattern Recogn. **47**(11), 3665–3680 (2014)
4. Burduk, R.: Classifier fusion with interval-valued weights. Pattern Recogn. Lett. **34**(14), 1623–1629 (2013)
5. Choraś, M., Kozik, R.: Machine learning techniques applied to detect cyber attacks on web applications. Logic J. IGPL **23**(1), 45–56 (2015)
6. Cohn, D., Atlas, L., Ladner, R.: Improving generalization with active learning. Mach. Learn. **15**(2), 201–221 (1994)
7. Cyganek, B.: One-class support vector ensembles for image segmentation and classification. J. Math. Imaging Vis. **42**(2–3), 103–117 (2012)
8. Forczmański, P., Łabędź, P.: Recognition of occluded faces based on multi-subspace classification. In: Saeed, K., Chaki, R., Cortesi, A., Wierzchoń, S. (eds.) CISIM 2013. LNCS, vol. 8104, pp. 148–157. Springer, Heidelberg (2013). doi:10.1007/978-3-642-40925-7_15

9. Frank, A., Asuncion, A.: UCI machine learning repository (2010)
10. Frejlichowski, D.: An algorithm for the automatic analysis of characters located on car license plates. In: Kamel, M., Campilho, A. (eds.) ICIAR 2013. LNCS, vol. 7950, pp. 774–781. Springer, Heidelberg (2013). doi:10.1007/978-3-642-39094-4_89
11. Freund, Y., Seung, H.S., Shamir, E., Tishby, N.: Selective sampling using the query by committee algorithm. Mach. Learn. **28**(2–3), 133–168 (1997)
12. Giacinto, G., Roli, F.: An approach to the automatic design of multiple classifier systems. Pattern Recogn. Lett. **22**, 25–33 (2001)
13. Kuncheva, L.I.: Combining Pattern Classifiers: Methods and Algorithms. Wiley, New Jersey (2004)
14. Rejer, I.: Genetic algorithm with aggressive mutation for feature selection in BCI feature space. Pattern Anal. Appl. **18**(3), 485–492 (2015)
15. Stefanowski, J., Pachocki, M.: Comparing performance of committee based approaches to active learning. In: Recent Advances in Intelligent Information Systems, pp. 457–470 (2009)

# The EOH Line Selector for Images with Downgraded Size for Mobile Robots Steering

Piotr Lech$^{(\boxtimes)}$ and Jarosław Fastowicz

Department of Signal Processing and Multimedia Engineering,
West Pomeranian University of Technology, Sikorskiego 37,
70–313 Szczecin, Poland
{piotr.lech,jaroslaw.fastowicz}@zut.edu.pl

**Abstract.** This paper presents the idea of algorithm for detecting lines having a defined direction in digital images based on selected Region of Interest containing those lines. The proposed algorithm uses data reduction in order to simplify information processing. The goal of the algorithm is the reduction of data size based on simple random sampling method. The Edge Oriented Histogram algorithm is used to designate the ROI blocks with information of potential places in image containing lines oriented in defined direction. This approach is proposed for low-computational power systems, embedded systems or video based control of mobile robots based on image analysis. The nearly real-time algorithm has been tested on real corridor image data sets obtained from a high resolution camera. The practical test implementation as a part of mobile robot steering algorithm is presented.

**Keywords:** Image analysis · EOH · ROI · Downgrading image resolution

## 1 Introduction

A typical solution to support the navigation in a confined space is the correct orientation according to the known solid objects placed in the test environment. A simplified representations of objects using their edges as stroke lines is often assumed. These lines are considered as natural landmarks in space. For example, in the case of corridors, landmarks are natural elements, such as doors, windows or other similar elements [1]. Image based line detection is a popular issue in mobile robot navigation. Algorithms for processing of high resolution images require significant computing power, so it is a problem in the case of full-resolution images in lower performance microprocessor systems. The critical point for line detection in the vision data processing flow is the point where the computer has to decide which pixels may represent the line. The line detection process can be greatly improved by reducing the amount of analyzed data. Gradual reduction of the information in the image processing algorithms is a commonly used approach, e.g. by conversion of color image to grayscale (or binary

© Springer International Publishing AG 2017
R.S. Choraś (ed.), *Image Processing and Communications Challenges 8*,
Advances in Intelligent Systems and Computing 525, DOI 10.1007/978-3-319-47274-4_15

black and white) or reduction of image size (resolution). For video sequences some background estimation and removal algorithms can also be used for the reduction of the amount of processed data [6].

Simultaneously with the reduction process, searching for information relevant to the specified issue (e.g. shape, size, orientation) takes place as well. This reduction sometimes causes an additional elimination of unnecessary and harmful information (noise, artifacts etc.). Reducing resolution is a way to reduce the amount of analyzed elements in the digital image, another way is the reduction of the analyzed area using a ROI filter such that the probability of the line detection is high. It is possible to use both methods simultaneously. In this paper the modified Edge Oriented Histogram (EOH) [10] is presented and applied in ROI filter algorithm. Some of popular algorithms used for line detection are based on edge detectors, such as Sobel or Canny filters, or applying pixel intensity based thresholding e.g. Otsu. Some algorithms based on thresholding or histogram methods can be converted to a fast method e.g. binarization with fast Otsu method [4,5]. The morphology and the median filter or the Hough Transform can also be used for selecting lines with assumed direction [7].

The EOH has been developed as a part of popular real-time algorithms e.g. SIFT, MPEG-7 [10,11]. However, in these cases, to achieve the parameters of the RT system, the potential of the hardware equipment is used, e.g. in the parallel processing. The developed nearly real-time algorithm is dedicated for microprocessors with low computational power.

## 2   Simple Random Sampling as a Method of Image Data Reduction

Widely available video cameras are often the source of high-resolution images. Such images cause some difficulties with their processing and storage. Therefore image resampling is a popular technique allowing to prepare a new version of original image based on mathematical formulas. However, the proposed method of downsampling is based on division into blocks and a statistical sampling inside each block, where new samples are drawn by a pseudo-random number generator with a uniform distribution.

The mean value of chosen pixels is then calculated. This process is illustrated on Fig. 1. The new image after downsampling is stored as a new low resolution image. The new matrix is an equivalent of the old image which can be easier for further processing using standard operations. It is possible to store a new calculated value in the vector, for example, it may be useful for histogram calculation. Sometimes, for blurry images such an operation can make it sharpen or slightly reduce noise. Proposed method is faster than a popular alternative method for scaling images based on known procedures like nearest-neighbor or bilinear interpolation [9].

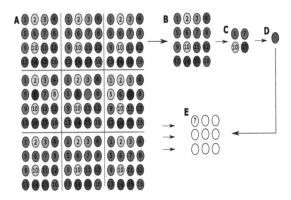

**Fig. 1.** Image downgrading process (A – original image divided into 9 parts, B – base for Simple Random Sampling, C – 4 drawn pixels, D – new calculated value of pixel, E – new downgraded image)

## 2.1   Comparison of Methods for Image Size Reduction

Test images have been reduced using the following methods:

1. method proposed in paper with block size $M \times M$ pixels and new value of luminance pixel has been calculated as the average value for $n$ drawn pixels,
2. nearest-neighbor interpolation with the same output image size,
3. bilinear interpolation with the same output image size.

The time required to complete the tasks has been measured and the results have been normalized assuming that the longest time is equal to 1 (Table 1).

**Table 1.** Comparison of the computation time

| Downgrading method for block | Normalized computation time [%] |
|---|---|
| $48 \times 48$ pixels average for $n = 16$ drawn pixels | 53 |
| $8 \times 8$ pixels average for $n = 4$ drawn pixels | 62 |
| Nearest-neighbor interpolation | 75 |
| Bilinear interpolation | 100 |

## 3   Extraction of Regions with Lines with Known Orientation

In the EOH algorithm the edge orientation is evaluated by searching the maximum response over an edge filter, directly calculating the gradient in each bin. The performance of the algorithm is dependent on the number of bins. The gradient magnitude and gradient orientation are calculated for each pixel. The

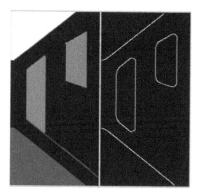

**Fig. 2.** Synthetic image (left – grayscale, right – binary after edge detection)

**Table 2.** Table of one block direction coded

| Line direction | Binary code |
|---|---|
| Horizontal | 1100 |
| Vertical | 0011 |
| 45 degree | 0110 |
| 135 degree | 1010 |
| Non direction | 0000 |

algorithm conducts the edge detection using Canny method by default which can be effectively applied due to the use of a relatively small image size. The gradient orientation value is divided into $K$ bins where $K = 5$ for five assumed directions is required with assumed Sobel filter orientation for vertical, horizontal, 45 degree, 135 degree and non-directional edges. At this point it is similar to the MPEG-7 "Edge Orientation Histogram" descriptor. The Edge Orientation Histograms in each orientation bin $k$ of cell $C_i$, $E_k(C_i)$, are calculated by summing all gradient magnitudes with orientations belonging to bin $k$. The set of $k$ EOH features for one cell can be expressed as the ratio of the bin value on single orientation to the sum of all bins values according to the formula:

$$EOH_{C_{i,k}} = \frac{E_k(C_i) + e}{\sum_{j=1}^{K} E_j(C_i) + e} \tag{1}$$

where $e$ is an additional value allowing correct calculations when denominator is close to zero.

The dominant value of bin for known orientation is minimal information necessary for preparation of the logical condition used for the choice of blocks containing lines. The dominant value has been defined as the maximum value stored in histogram bins. For each block dominant direction is coded (Table 2) and stored in two dimensional matrix. The matrix is necessary for preparation of

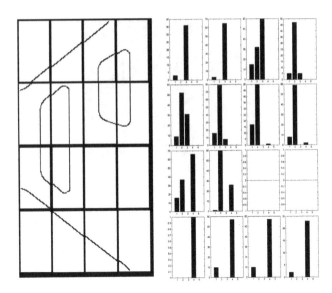

**Fig. 3.** The idea of region selection algorithm (left – negative of blocks, right – EOH for each block)

the ROI filter. The analysis of the neighborhood of these coded direction values is the base for selection of potential places of line occurrence for the specified direction. It has been found that for long lines for each single block the trend of a certain direction is found to be very similar also in their closest neighborhood (examples using a synthetic image are shown on Figs. 2 and 3).

## 4    Tests

The steps of test algorithm are:

1. the high resolution image is downgraded by dividing into blocks and simple random sampling method is used,
2. new image is divided into $r^2$ blocks (where $r$ is individually declared value for images)
3. for each block Edge Oriented Histogram is calculated,
4. for each block the code of direction is calculated and stored in two dimensional matrix,
5. the ROI is calculated based on neighborhood coded direction stored in two dimensional matrix,
6. all blocks except those with the desirable directions of the lines are removed.

### 4.1    Example of ROI Selection

In the presented example, the ROI filter has the task of selecting the lines according to the selected direction which have at least two neighbors. The application of the ROI selector for a corridor line is presented in Fig. 4.

**Fig. 4.** Example of ROI based corridor line extractor (downscaled image, Canny edge, vertical lines, 45 and 135 degree lines)

**Fig. 5.** Example of ROI based roadsides selector

In the second example the task is to select only those lines which are arranged in a direction of 45 or 135 degrees. The sample drawing is done selecting a line painted on the roadway and its roadsides. ROI is defined for five neighbors with the same coded directions. In this example the traffic lane and right roadside are properly extracted but the left roadside has been omitted (Fig. 5).

## 4.2  Example of Mobile Robot Steering

The task of mobile robot control is related to maintaining the forward direction of its motion, keeping the same distance from the walls of the corridor. Controlling the robot using the proposed idea is presented in Fig. 6. The proposed control algorithm has to guarantee the continuity of the line for its proper work but in the case of loss of continuity of the detected line the previously calculated control will be continued. Plane of images obtained from the camera in the 3D space must always be perpendicular to the axis of movement defined by the parallel walls of the corridor for the proper designation of the deviation from the central point of robot position (which belongs to this axis). It is important that the measurement is always made at the level of the selected default horizontal line localized identically on each of images. The test line is arranged so that the divided image proportion is 1:5 and is placed closer to the bottom of the image. The tests results have been obtained with processing speed equal to 12 video frames per second.

The algorithm has been checked in off-line mode, including only the selected points from the route of the mobile robot for which the images have been captured and the distance measurements from the optical axis of the robot's camera to the walls have been made using a hand laser range finder.

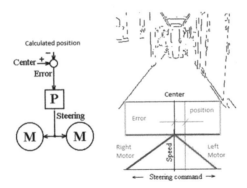

**Fig. 6.** Vision based differential steering idea of a mobile robot

**Table 3.** Example of calculated error

| Deviation | Vision method's relative error | Laser range finder's relative error |
|-----------|-------------------------------|-------------------------------------|
| 10 [cm]   | 8 %                           | 6 %                                 |
| 50 [cm]   | 7 %                           | 5 %                                 |
| 100 [cm]  | 8 %                           | 4 %                                 |
| 150 [cm]  | 9 %                           | 4 %                                 |

The results of the measurements were relative to the reference ones obtained on the basis of designated by the range finder for the saved images. The results are shown in Table 3, for both methods the errors relative to the reference measurements are similar.

## 5    Conclusions

The proposed ROI filter for a line selector with assumed direction based on Edge Oriented Histogram and downgrading image resolution by simple random sampling is a useful solution for algorithms requiring simple implementation and low computational complexity. It has been shown during the tests that the selection criterion of a line needs to be developed, due to some problems with the correct selection in the case of a large number of lines directed in different directions. Despite the found flaws, the experiments have confirmed the usefulness of the developed algorithm for the control of a mobile robot. Further work will be related to the use of advanced compression techniques [3] to improve the quality of detection line.

# References

1. Cui, X.N., Kim, Y.G., Kim, H.: Floor segmentation by computing plane normals from image motion fields for visual navigation. Int. J. Control Autom. Syst. **7**(5), 788–798 (2009). ISO 690
2. Fazl-Ersi, E., Tsotsos, J.K.: Region classification for robust floor detection in indoor environments. In: Kamel, M., Campilho, A. (eds.) ICIAR 2009. LNCS, vol. 5627, pp. 717–726. Springer, Heidelberg (2009). doi:10.1007/978-3-642-02611-9_71
3. Forczmański, P., Maleika, W.: Near-lossless PCA-based compression of seabed surface with prediction. In: Kamel, M., Campilho, A. (eds.) ICIAR 2015. LNCS, vol. 9164, pp. 119–128. Springer, Heidelberg (2015). doi:10.1007/978-3-319-20801-5_13
4. Lech, P., Okarma, K.: Optimization of the fast image binarization method based on the Monte Carlo approach. Elektronika Ir Elektrotechnika **20**(4), 63–66 (2014)
5. Lech, P., Okarma, K., Tecław, M.: A fast histogram estimation based on the Monte Carlo method for image binarization. In: Choras, R.S. (ed.) Image Processing and Communications Challenges 5, pp. 73–80. Springer International Publishing, Switzerland (2014)
6. Mazurek, P., Okarma, K.: Background suppression for video vehicle tracking systems with moving cameras using camera motion estimation. In: Mikulski, J. (ed.) TST 2012. CCIS, vol. 329, pp. 372–379. Springer, Heidelberg (2012). doi:10.1007/978-3-642-34050-5_42
7. McDonald, J.B., Mc Donald, J.B., Shorten, R., Franz, J.: Application of the hough transform to lane detection in motorway driving scenarios (2001)
8. Okarma, K., Lech, P.: A fast image analysis technique for the line tracking robots. In: Rutkowski, L., Scherer, R., Tadeusiewicz, R., Zadeh, L.A., Zurada, J.M. (eds.) ICAISC 2010. LNCS (LNAI), vol. 6114, pp. 329–336. Springer, Heidelberg (2010). doi:10.1007/978-3-642-13232-2_40
9. Suzuki, K., Saito, S., Chang, Y., Nakajima, M.: Downscaling for images having thin line structure. In: Poster Session Presented at: The 5th International Symposium on Non-photorealistic Animation and Rendering (2007)
10. Vinay, A., Kathiresan, G., Mundroy, D.A., Nandan, H.N., Sureka, C., Murthy, K.B., Natarajan, S.: Face recognition using filtered eoh-sift. Procedia Comput. Sci. **79**, 543–552 (2016)
11. Vo, D.M., Zell, A.: Real-time face recognition using local ternary patterns with collaborative representation-based classification for mobile robots. In: Menegatti, E., Michael, N., Berns, K., Yamaguchi, H. (eds.) Intelligent Autonomous Systems 13. AISC, vol. 302, pp. 781–793. Springer, Heidelberg (2016). doi:10.1007/978-3-319-08338-4_56

# Swipe-Like Text Entry by Head Movements and a Single Row Keyboard

Adam Nowosielski[✉]

Faculty of Computer Science and Information Technology,
West Pomeranian University of Technology,
Żołnierska 52, 71-210 Szczecin, Poland
anowosielski@wi.zut.edu.pl

**Abstract.** Swipe typing on haptic devices is gaining increasing attention among developers and popularity among users. By sliding a finger or stylus through consecutive letters, without any form of selection (e.g. pressing), it offers substantial acceleration and ease of writing. In this paper, an adaptation of swipe typing technique for touchless environment operated by head movements is proposed. The head movements enable directional manipulation and the swipe-like typing procedure solves the problem of the key press. Such interface is generally designed for physically challenged people who are unable to use standard input devices (mouse and keyboard).

## 1 Introduction

Traditional typing using the physical keyboard requires the user to press each character composing a word. The same situation is encountered during traditional (without any improvements) typing on soft keyboards presented on the screen: every key, representing a single character, has to be separately pressed. Different principles, on the other hand, form the foundation of the swipe typing: the entire word is entered by a single gesture. The user usually touches the key representing the first letter of a word, then drags through subsequent characters to finally raise the pointer (finger or stylus) over the last letter. A word is entered using a single and continuous gesture.

In swipe typing techniques a path of the performed gesture consists of many redundant characters. In order to determine the typed phrase this path must be analyzed using a word search engine. There are many proprietary solutions of swipe typing available on the market. All of them, however, focus on touchscreen devices.

Few examples presented in the literature confirm that the adaptation of swipe typing technique to touchless environment is possible and can improve the overall interaction [3,5,13]. In [3] the user's hand and fingers are tracked in relatively close distance from the screen. The path made by hand movements is analyzed for the intended word. The separation of words is achieved with a pinch gesture. Another solution is proposed in [13] where interface components are controlled by the Kinect-based hands motion tracking module. The clench fist gesture is

© Springer International Publishing AG 2017
R.S. Choraś (ed.), *Image Processing and Communications Challenges 8*,
Advances in Intelligent Systems and Computing 525, DOI 10.1007/978-3-319-47274-4_16

used for the separation of words. The velocity of movement through the keyboard is analyzed for the collection of suitable characters.

In this paper, a proposition of swipe typing technique adaptation for touchless environment operated by head movements is presented. Such interface is generally designed for physically challenged people who are unable to use standard input devices (mouse and keyboard).

The rest of the paper is structured as follows. In Sect. 2 related works are presented and discussed. Section 3 introduces the concept of swipe typing in head operated interface. Its details are provided in Sect. 4. Final conclusions and a summary are provided in Sect. 5.

## 2 Related Works

Most head operated interfaces focus on conventional mouse replacement and the pointer manipulation in the graphical user interface. The problem of touchless typing is neglected in the contemporary research. Text entry, however, is the tasks in human-computer interaction equally important as the mouse input. Although alternatives to touchless text typing exist (e.g. speech recognition, eyetracking) in this paper the emphasis is placed on computer vision techniques. Vision-based text entry interfaces are still rare and insufficiently studied [2]. They are also considered as a novel issue [3,6].

### 2.1 Head Operated Interfaces

Head operated interfaces can be defined as touchless interaction interfaces where manipulation is achieved by a form of the head movements. These interfaces may also focus directly on a face or facial features. Some solutions use markers attached to distinctive parts of the head (e.g. middle of the forehead). Their aim is to simplify the processes of detection and tracking. The detection and tracking, in fact, form the basis for the interface.

An example of mouse replacement with the head movements is SmartNav from NaturalPoint (http://www.naturalpoint.com/smartnav/). It is a commercial product that offers hands-free cursor control also with movements of other part of the body. This is a typical marker-based solution. It works in infrared light and uses: infrared emitters, reflector and an infrared camera.

Another proposition for hands-free mouse alternative with head movements is the Enable Viacam (eViacam, http://eviacam.sourceforge.net). It is an open source solution that works in visible spectrum without the need of additional markers. It works on standard PC equipped with a webcam. Face and mouth detection and subsequent tracking are of most importance here.

Two solutions presented above are examples of finished products. There are also some propositions in the scientific literature. In [4] the position of cursor is controlled by relative position of the nostrils to the face region. The skin colour method is used first for the face detection. The nostrils are then found using heuristic rules.

In [11] the mouse control is achieved with eye and nose tracking. The eye regions are processed for recognition of winks (used as the mouse clicks). The user's face is detected initially by means of the well known Viola and Jones algorithm [12]. Then, a Gaussian model in RGB is used to represent the skin colour probability density function. Another solution that uses the image plane position of the eyes is presented in [8]. The changes in the eyes localization are transferred to a control system. Eye blink is used for confirmations. The involuntary blink and steering blink are distinguished on the basis of the closure duration.

[10] proposes touchless interface using multiple facial feature detection and tracking. Mouse movements are implemented based on the user's eye movements. Clicking events, on the other hand, are implemented using the user's mouth shapes (opening/closing). Face detection is achieved here by a skin-color model and connected-component analysis. The eye regions are localized by a neural network texture classifier. The mouth region is localized using an edge detector. After the detection, the eye regions are tracked with a mean-shift algorithm. The mouth regions are tracked using template matching. Authors of the solution also provided a "spelling board" which substitute for a standard keyboard (operated with eye movements and mouth shapes).

Touchless typing with head movements is also proposed in [2]. Here, the head movements are supported with three face gestures chosen for a key selection: mouth open, brows up and brows down.

## 2.2  User's Head Detection for Touchless Interfaces

The main tasks in head operated interfaces are the head/face detection and tracking. Considering the visible spectrum without the use of additional markers, the user's face in front of the screen can be detected and tracked using the following computer vision techniques:

– continuous face detection (i.e. tracking by detection),
– face detection using individual frame and tracking in subsequent frames,
– stereo vision,
– background modeling.

The continuous face detection is reserved for fast methods. Those methods frequently offer approximate positions and are prone to errors. More accurate approaches are slower and not sufficient for the real-time interaction. They are, however, willingly combined with faster tracking algorithms. In stereo vision the scene is observed using a pair of calibrated cameras or a single stereoscopic camera. Individual images are combined to form a depth map which can be easly analyzed for the presence of the user silhouette, upper body or a head. Background modeling approaches also offer a robust method for user detection. They are, in turn, sensitive for changes in the scene background and are not suited for dynamic surroundings.

## 3   Interface Concept

The first assumption for the interface is operation with head movements. Head movements should be simple and straightforward, easy to understand, learn and operate by new users. It is crucial to minimize the number of different movements. To achieve the presented goal the single row alphabetical keyboard is used. The alphabetical order is intuitive and easily operable. With a single row layout the selection of a letter requires only movements to the left and right. By adopting the swipe typing technique there is no need for a special gesture corresponding to the key press.

A characteristic feature of the single-row keyboard is frequent change of the movement direction (see the example in Fig. 1). The change of the movement direction occurs on the intended letters. A slowdown occurs and the interaction with the key is noticeably extended. It is important temporal characteristic. In the second case, when the intended letter is located between direction changes (i.e. the intended letter occurs on the path) the user is expected to slow the movement in the area of that letter. Such situation appears in the exemplary "text" word (Fig. 1) and the second "t" letter placed between "x" and the space.

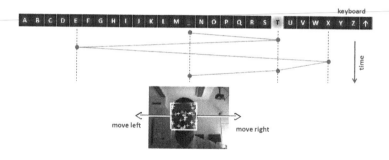

**Fig. 1.** Interface concept: the exemplary interaction during "text" word entry

At the beginning of the interaction the central part of the keyboard is highlighted. This region is special and forms a neutral area. Any slight movements of the head are used for the adaptation of the reference central head position (necessary in order not to immobilize the user; the user is allowed to make free movements on a limited extent). This region is also assigned for the space key. With a greater head movement the swipe procedure is launched. The user moves his head horizontally with the aim to reach subsequent characters of the intended word. During the movement the appropriate letters are highlighted according to the mapping procedure from the head position. The absolute measure of displacement is used, i.e. the distance of the head shift corresponds to a particular letter. Such an approach is faster compared to relative displacement where head shifts result in scrolling through consecutive letters. Moreover, eventual inaccuracies can be eliminated in the next stages.

The typing starts from the central region and finishes in that region too. During the movements all characters crossed over form a sequence. This sequence consist of many redundant and unintended letters. The sequence has to be analyzed and only meaningful letters should remain. There are many combinations which are subjected to dictionary analysis. The resultant correct words are presented for the user as suggestions.

Based on the concepts presented above a prototype interface have been created. It was evaluated by typing separated words. The checking procedure consisted of time measurements and verifications whether the intended words occurred among returned suggestions. In the final product, after the swipe gesture is made, the suggested words might be displayed in another row with the most suited word highlighted. The transition between words could be achieved with left and right head movements. The switch between rows could be arranged with head up and down tilt gesture. Those gestures could also be used for individual letters entry, when no suggestion exist (for unique words).

## 4   Development of the Interface

Development of the interface requires the adequate techniques for user's head detection and tracking. Secondly, methods for path analysis and dictionary support have to be proposed. Both issues are addressed below.

### 4.1   Detection and Tracking of the User's Head

In the proposed interface for the initial face detection task the well known and widely used Viola and Jones approach [12] has been used. The algorithm offers good results even for complex scenes. In the context of human-computer interaction, when the user is present in front of the screen, few frames are available, and the lighting conditions are at least moderate, the results are perfect. Although the tracking by detection is possible using only the Viola and Jones approach, problems with efficiency may occur on slower hardware configurations.

The Viola and Jones approach is frequently assisted with the Kanade-Lucas-Tomasi (KLT) feature-tracking procedure. The algorithm, also used in the proposed interface, has the following steps. First, initial points for tracking are selected. There are many possibilities for the selection of initial points: corners using minimum eigenvalue algorithm [9] (used in the proposed interface), corners using FAST algorithm [7], SURF features [1]. In the second stage, for each point the tracker attempts to find the corresponding point in new frame. The algorithm is stable for rigid objects that do not change shape. For a face, during the tracking, some points can be lost. The procedure of new points selection is however fast and imperceptible for the user. It can be launched when too many points are lost or periodically when the head is over the neutral area (between consecutive words).

The referred approach offers stable and reliable head position for the real-time interaction. The center of the bounding box around the face is used for the steering purposes - detection and range of the horizontal head movements.

## 4.2   Swipe Gesture Analysis

The crucial part of the developed interface is the approach to swipe gesture analysis and word matching. While the user moves his head above the virtual keyboard the appropriate letters are highlighted. The mapping is absolute: the distance of the head shift corresponds to a particular letter. During the calibration step it is the user who chooses convenient ranges of the movements. The extreme shifts corresponds to most distant characters on the keyboard.

On the head movements the trajectory (consisting of 1D $x$ coordinates) and temporal characteristics (duration of the interaction with subsequent keys, in milliseconds) are recorded for further path analysis. Initially, all characters located on the performed path are selected. Temporal characteristics are used to remove characters with the shortest time of focus. Figure 2 presents three examples for the word "computer". The abscissa represents the interaction with subsequent characters on the keyboard. The ordinate shows the function of the time of this interaction (the inverse of the square of the time is used as the measure). For a single word the mean value for the function is calculated (plotted as a red dashed line). The threshold was set as 0.15 of this value (plotted as a red solid line). Only the characters below the threshold are recognized as intended (indicated by a blue square). Frequently, intended letters form connected groups containing unnecessary characters.

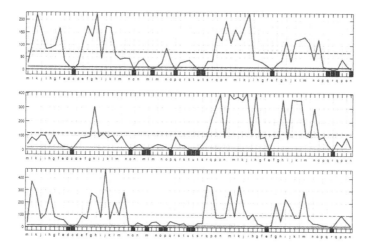

**Fig. 2.** Temporal characteristics for three exemplary interactions during the "computer" word entry (Color figure online)

The procedure of the word matching is two-staged. First, from the intended letters a string is formulated. It is dictionary matched using the external spell checker (the proprietary Microsoft Word spell checker was engaged). Because the input string might be noisy the typical word processor spell checker often fails. For the examples from the Fig. 2 strings are respectively: "colputeqrqn", "comlptuter", "dcomoputer". The spell checker fails with the first string. With two others, the correct suggestions are provided.

In the second step new possible words are generated and dictionary checked for verification. Two approaches have been considered. In the first case, new strings are generated substituting original letters in the initial string with alphabetical neighbors (e.g. "l" is substituted with "k" and "m") and with omissions of string letters. In the second approach, new strings are formulated by selection only individual letters from connected groups of intended letters. Omissions of string letters and connected groups are also employed.

Two approaches applied in the second step work differently. The first is better with longer words and the second when the user's head movements are hesitant. The best solution, however, is the combination of both approaches.

The evaluation of the interface was conducted on typing separated words. The checking procedure consisted of time measurements and verifications whether the intended words occurred among the returned suggestions. For the acquainted users the average interaction time per character was 1.78 s which gives the value of 33.7 cpm (chars per minute). This value, however, does not take into account: necessary breaks between words during continuous writing, time of the appropriate word selection from provided suggestions, and finally the time of suggestions generation. In the prepared prototype, the last one was the most critical. With generation of many combinations (the allowance of many changes) the procedure took even several seconds. This was the reason for the separated words evaluation procedure. The results, however, were encouraging. They prove that swipe-like text entry by head movements using a single row keyboard is possible and can be effective considering the interaction time.

## 5   Conclusions

In the paper, the problem of text entry using head movements have been addressed and the appropriate interface proposed. The novel element of the interface is the adaptation of swipe typing technique to a single row keyboard operated with horizontal head movements. Initial experiments carried out on developed prototype demonstrated that such an approach is possible and allows convenient touchless typing. There is no need for a press gesture - a slow down of the movement is sufficient - hence the proposed swipe-like name for a technique.

The interface of touchless text entry is of particular importance for physically challenged people who are unable to operate the standard computer input devices. Since most non-contact interfaces for typing focus on conventional mouse replacement the touchless text entry still remains a challenge.

# References

1. Bay, H., Ess, A., Tuytelaars, T., Gool, L.V.: Speeded-up robust features (SURF). Comput. Vis. Image Underst. **110**(3), 346–359 (2008)
2. Gizatdinova, Y., Spakov, O., Surakka, V.: Face typing: vision-based perceptual interface for hands-free text entry with a scrollable virtual keyboard. In: IEEE Workshop on Applications of Computer Vision 2012, Breckenridge, CO, USA, pp. 81–87 (2012)
3. Markussen, A., Jakobsen, M.R., Hornbæk, K.: Vulture: a mid-air word-gesture keyboard. In: Proceedings of the SIGCHI Conference on Human Factors in Computing Systems, CHI 2014, pp. 1073–1082 (2014)
4. Morris, T., Chauhan, V.: Facial feature tracking for cursor control. J. Netw. Comput. Appl. **29**(2006), 62–80 (2006)
5. Nowosielski, A.: Evaluation of touchless typing techniques with hand movement. In: Burduk, R., Jackowski, K., Kurzyński, M., Woźniak, M., Żołnierek, A. (eds.) Proceedings of the 9th International Conference on Computer Recognition Systems CORES 2015. Advances in Intelligent Systems and Computing, vol. 403, pp. 441–449. Springer, Switzerland (2016)
6. Nowosielski, A.: QWERTY- and *8pen*- based touchless text input with hand movement. In: Chmielewski, L.J., Kozera, R., Shin, B.-S., Wojciechowski, K. (eds.) ICCVG 2014. LNCS, vol. 8671, pp. 470–477. Springer, Heidelberg (2014). doi:10. 1007/978-3-319-11331-9_56
7. Rosten, E., Drummond, T.: Fusing points and lines for high performance tracking. In: Tenth IEEE International Conference on Computer Vision (ICCV 2005), vol. 2, pp. 1508–1511 (2005)
8. Santis, A., Iacoviello, D.: Robust real time eye tracking for computer interface for disabled people. Comput. Methods Programs Biomed. **96**(1), 1–11 (2009)
9. Shi, J., Tomasi, C.: Good features to track. In: Proceedings CVPR 1994, pp. 593–600 (1994)
10. Shin, Y., Ju, J.S., Kim, E.Y.: Welfare interface implementation using multiple facial features tracking for the disabled people. Pattern Recogn. Lett. **29**(2008), 1784–1796 (2008)
11. Varona, J., Manresa-Yee, C., Perales, F.J.: Hands-free vision-based interface for computer accessibility. J. Netw. Comput. Appl. **31**(4), 357–374 (2008)
12. Viola, P., Jones, M.: Robust real-time face detection. Int. J. Comput. Vision **57**(2), 137–154 (2004)
13. Wierzchowski, M., Nowosielski, A.: Swipe text input for touchless interfaces. In: Burduk, R., Jackowski, K., Kurzyński, M., Woźniak, M., Żołnierek, A. (eds.) Proceedings of the 9th International Conference on Computer Recognition Systems CORES 2015. Advances in Intelligent Systems and Computing, vol. 403, pp. 619–629. Springer, Switzerland (2016)

# The Feature Extraction From
# the Parameter Space

Adam Marchewka[(✉)] and Mirosław Miciak

Institute of Telecommunications and Computer Science,
Faculty of Telecommunication, Information Technology and Electrical Engineering,
University of Technology and Life Sciences (UTP), Kaliskiego 7,
85-789 Bydgoszcz, Poland
{adimar,miciak}@utp.edu.pl

**Abstract.** In this article a new solution of characteristic points of the
image from the parameter space is presented. The implemented algorithm can be applied to recognition of postcode information. The main
objective of this article is to use the Radon Transformation parameter
space to obtain an invariant set of character image features, on basis of
which unknown digits can be classified. The reported experiments results
prove the effectiveness of our method.

## 1 Introduction

The postal addresses most often written in the form of a postal code consists of
set of numbers and can be presented in the form of images representing individual characters. The character images contain noises and interferences, moreover
image may be subjected to many transformations such as: rotation and scale
change. This situation makes it difficult direct determination of the features
of the image character. In connection with the above, the authors propose a
transformation of the image to the parameter space, where you will be able to
carry out the operation of normalization and correction of rotation, and also you
will be able image processing on a scale of gray and character images with the
interferences.

## 2 Using the Data of Parameter Space in the Task of Identifying Postal Address Information

The zip-code character images data are presented in the form of a parametric
space from Radon transformation. The parametric space contains relevant information relating to the description of the zip-code character. This allows to obtain
a number of features enabling assigned unknown character to the appropriate
class. The parametric representation of the elements of character (eg. a straight
line) is represented in the form of local maxima of accumulator. It allows to
describe a character. The analysis of the parametric images from own database
allow to observe of similar distributions (in terms of the maximum number of

R.S. Choraś (ed.), *Image Processing and Communications Challenges 8*,
Advances in Intelligent Systems and Computing 525, DOI 10.1007/978-3-319-47274-4_17

values) within each class of characters. It is possible to consider the number and location of decomposition peaks parametric representation as a feature of the character image.

## 2.1   Radon Transformation

In recent times the parametric transformation - Radon Transformation have taken much more attention. This opreation is able to change two dimensional images with lines into a domain of possible line parameters, where each line in the transformed image will give a peak placed at the issuable line parameters. This has lead to many line detection applications [6]. The Radon Transformation is a basic tool which is used in miscellaneous applications such as radar imaging, geophysical imaging, nondestructive testing tool and medical imaging tool [7]. The Radon Transform can calculate projections of an image data along strictly defined directions. A projection of a two dimensional function $f(x, y)$ is a set of line integrals along an image input array. The Radon formula calculates the line integrals from a number of sources along parallel paths, or beams, in a fix sets direction. For example the beams are spaced one point unit separately. To represent whole calculated image, the Radon function takes many parallel beam projections of the image matrix from various angles for example by rotating image array around its centre point. The Fig. 1 shows a single sample projection at a defined rotation angle.

The Radon Transformation is the set of projection of the image matrix intensity along a radial line rotated at a perpendicular angle. In this manner the new radial coordinates are the values along the $x'$-axis, which is oriented at specific $\theta$ degrees counter clockwise from the $x$-axis. Whereas the center of both axes is the center point of the transformed image. Therefore, the line integral

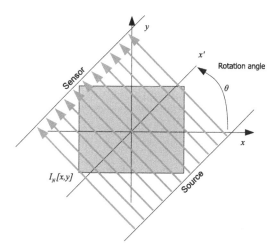

**Fig. 1.** The sample projection at a defined angle

of $f(x,y)$ in the vertical direction is the Radon function projection of $f(x,y)$ onto the $x$-axis, whereas the line integral in the horizontal direction is the projection function of $f(x,y)$ onto the $y$-axis. Accordingly, the projections can be defined along any specific angle $\theta$, using general form of equation of the Radon Transformation [1,2]:

$$R_\theta(x') = \int_{-\infty}^{\infty} \int_{-\infty}^{\infty} f(x,y)\delta(x\cos\theta + y\sin\theta - x')dxdy \qquad (1)$$

where:

$$\delta(x\cos\theta + y\sin\theta - x') \qquad (2)$$

where *delta* is the delta function with not zero value only for one argument equal 0, and:

$$x' = x\cos\theta + y\sin\theta \qquad (3)$$

$x'$ is the simple perpendicular distance of the beam from the center and $\theta$ is the angle of range of the beams.

One of the most important properties of the Radon Transform is the ability to detect lines (curves in general form of RT equation) from noisy images. Moreover, In addition, we find that Radon Transformation has certain interesting properties referring to the application of affine image transformations.

## 2.2 Extracting the Characteristic Points of the Parameter Space

As proposed by the authors and in the articles [4,5], Basic information received from the parameter representation space is a local maxima, which correspond to points of intersection lines (or their extensions) corresponding the shape of the character. Using the local maxima information of the parameter space, we can create representations of the character image. Additional information includes: co-ordinates and the value of the peaks in the parameter space. In this way, it is possible formulation of feature vector for the image of the character (Fig. 2).

The size of the parametric representation of the Radon transform are specified by means of parameters $\rho, \theta$. Fixing the angle step rotation of changes to every one degree will get 180 column parametric representations. Whereas, taking into account the image size is $m \times m$, the unit changes the length of the radius $\rho$ and the following relationship:

$$\rho = \left\lceil \frac{m}{2}\sqrt{2} \right\rceil \qquad (4)$$

this range includes changes in the length of the radius in the interval

$$(-\rho, .., 0, ...\rho) \qquad (5)$$

eg. for a image with dimensions $128 \times 128$ we get 183 lines representation of the Radon transform.

Considering the accumulator arrays $\tilde{A}_{cu}(\rho, \theta)$ are referred to the following parameters:

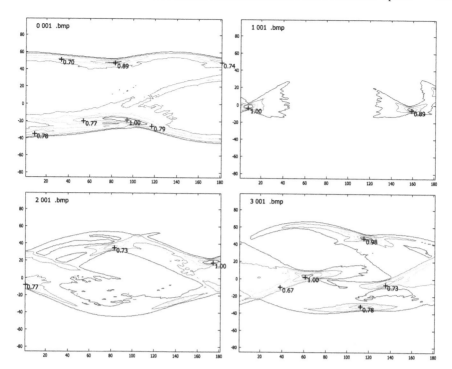

**Fig. 2.** Distribution of the maximum values for the selected characters

- the number of peaks of Accumulator array

$$lmax \tag{6}$$

- set of peak values of Accumulator array

$$\{wm_1, ..., wm_{lmax}\} \tag{7}$$

- set of coordinates of peaks of Accumulator array

$$\{(\theta_1, \rho_1)wm_1, ..., (\theta_{lmax}, \rho_{lmax})wm_{lmax}\} \tag{8}$$

Based on the analysis (Table 1) can be stated that the number of peaks is so diverse that they can not be a criterion for classification. In connection with the above determine is necessary for each maximum corresponding value $wm$ parametric representation and location coordinates $wm_\theta, wm_\rho$. The analysis results for own database of images of characters are presented in the Table 1. Based on the above feature vector characteristic points (CPC) could be written as:

$$FV_{CPC} = \left\{ lmax, ((w_{m_1}, w_{m\theta_1}, w_{m\rho_1}), ..., (w_{m_{lmax}}, w_{m\theta_{lmax}}, w_{m\rho_{lmax}})) \right\} \tag{9}$$

The feature vector for each character consists of two parts. The first part contains a number of peaks $lmax$, the second holds the value of $w_m$ and co-ordinates

**Table 1.** Compartments number of local maxims

| Digits | 0 | 1 | 2 | 3 | 4 | 5 | 6 | 7 | 8 | 9 |
|---|---|---|---|---|---|---|---|---|---|---|
| $lmax$ | 4÷10 | 1÷4 | 2÷5 | 3÷10 | 3÷8 | 3÷9 | 2÷10 | 2÷4 | 3÷10 | 3÷9 |

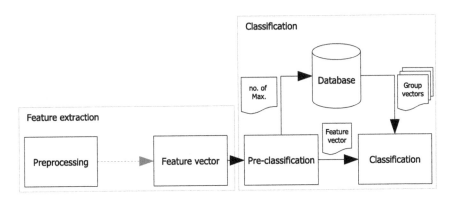

**Fig. 3.** The pre-classification scheme

$w_{m\theta}, w_{m\rho}$. The separation of the vector into two parts enables performing classification without engagement, based eg. On the number of maximum points.

The pre-classification scheme shown on Fig. 3 is based on the analysis of the number of local maxima obtained in the process of creating vector features. On the basis of this parameter is determined by a group of vectors from the database, which should be compared with the vector of the unknown character. The length of the feature vector (the largest number of points of local peaks) was determined on the basis of research using own database characters.

Based on the above, in the block of the pre-classification is created group of characters where the group number corresponds to the number of local peaks. Sample parameters vector features of test set are presented in the Table 2.

**Table 2.** The parameters of vector features $FV_{CPC}$

| Char | $l_{max}$ | $w_{m\theta_1}$ | $w_{m\rho_1}$ | $w_{m_2}$ | $w_{m\theta_2}$ | $w_{m\rho_2}$ | $w_{m_3}$ | $w_{m\theta_3}$ | $w_{m\rho_3}$ | $w_{m_4}$ | $w_{m\theta_4}$ | $w_{m\rho_4}$ | $w_{m_5}$ | $w_{m\theta_5}$ |
|---|---|---|---|---|---|---|---|---|---|---|---|---|---|---|
| 0 | 5 | 121 | 68 | 0,9 | 110 | 134 | 0,8 | 35 | 52 | 0,5 | 2 | 137 | 0,5 | 16 |
| 1 | 2 | 101 | 83 | 0,9 | 71 | 105 | | | | | | | | |
| 2 | 4 | 120 | 81 | 0,7 | 29 | 64 | 0,6 | 128 | 112 | 0,6 | 87 | 126 | | |
| 3 | 4 | 56 | 97 | 1,0 | 110 | 52 | 0,8 | 107 | 131 | 0,7 | 130 | 106 | | |
| 4 | 4 | 30 | 109 | 1,0 | 36 | 82 | 0,9 | 150 | 104 | 0,6 | 53 | 91 | | |
| 5 | 6 | 85 | 121 | 0,8 | 58 | 84 | 0,8 | 91 | 58 | 0,7 | 124 | 113 | 0,6 | 152 |
| 6 | 2 | 132 | 108 | 0,5 | 96 | 94 | | | | | | | | |
| 7 | 3 | 130 | 87 | 0,8 | 77 | 81 | 0,7 | 91 | 117 | | | | | |
| 8 | 3 | 65 | 82 | 0,9 | 147 | 100 | 0,5 | 20 | 97 | | | | | |
| 9 | 3 | 59 | 88 | 0,7 | 138 | 61 | 0,7 | 109 | 142 | | | | | |

## 2.3   Extracting of Connected Areas of the Parameter Space

In the second method, a feature vector is determined on the basis of the connected areas (BLOB). This method is based on determining the geometric parameters of separate areas of representation of the Radon transform. Considering the accumulator array $\tilde{A}_{cu}(\rho, \theta)$, for which the features are determined in the following way:

1. Determined are the limits of the areas on the basis of coordinate local maxima $w_{m\theta}, w_{m\rho}$ using thresholding matrix elements $\tilde{A}_{cu}$:

$$\tilde{\tilde{A}}_{cu}(\rho, \theta) = \begin{cases} \tilde{A}_{cu}(\rho, \theta) & \tilde{A}_{cu}(\rho, \theta) \geq thr \\ 0 & \tilde{A}_{cu}(\rho, \theta) < thr \end{cases} \tag{10}$$

where the threshold value of $thr$ is determined by:

$$thr = \frac{max\{w_{m_1}, ..., w_{m_{lmax}}\}}{2} \tag{11}$$

Thus there was obtained $lb$ areas $B = \{b_1, b_2, ..., b_{lb}\}$, which are then processed in a rectangular matrix containing the data of the accumulator belonging only to a selected area (Fig. 4):

$$(\rho_{max_k}, \theta_{max_k}) \tag{12}$$

and

$$(\rho_{min_k}, \theta_{min_k}) \tag{13}$$

for $k = 1, 2, .., lb$.

2. Then it is carried parameterization of elements from the set $B$. Selected values are as follows:
   - the local maximum value for the extracted area - $wm(b)$,
   - coordinates of the local maximum value for the extracted area - $wm_\theta(b)$, $wm_\rho(b)$,
   - amount of pixels for the extracted area - $fb(b)$, defined as:

$$fb(b) = \sum_\rho \sum_\theta \tilde{A}_{cu}(\rho, \theta) \tag{14}$$

   where

$$\tilde{A}_{cu}(\rho, \theta) = \begin{cases} 1 & \tilde{A}_{cu}(\rho, \theta) \geq thr \\ 0 & \tilde{A}_{cu}(\rho, \theta) < thr \end{cases} \tag{15}$$

   - angle $\alpha(b)$[1] from $Og$ to $\rho = 0$ [3],
   - coordinates of the centroid for the extracted area - $ws_\theta(b)$, $ws_\rho(b)$, defined as:

$$ws_\rho(b) = \frac{1}{fb(b)} \sum_\rho \sum_\theta \hat{A}_{cu}(\rho, \theta) \tag{16}$$

---

[1] The angle ranges $(-90, 90)$.

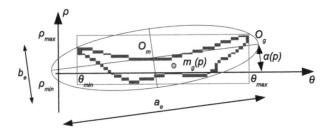

**Fig. 4.** The parameterization for selected areas: an area for image character *201.bmp*

where

$$\hat{A}_{cu}(\rho, \theta) = \begin{cases} \rho & \tilde{A}_{cu}(\rho, \theta) \geq thr \\ 0 & \tilde{A}_{cu}(\rho, \theta) < thr \end{cases} \tag{17}$$

for calculating $ws_\theta(b)$ consider condition:

$$\hat{A}_{cu}(\rho, \theta) = \begin{cases} \theta & \tilde{A}_{cu}(\rho, \theta) \geq thr \\ 0 & \tilde{A}_{cu}(\rho, \theta) < thr \end{cases} \tag{18}$$

– the main factor axis $Og$ and minor the axis $Om$ for the extracted area - $e_b$, defined as:

$$e_b = \sqrt{1 - \frac{b_e^2}{a_e^2}} \tag{19}$$

where: $a_e$ i $b_e$ - respectively, main axis length and minor axis length for the extracted area[2].

3. I've created a feature vector for the extracted areas (BLOB) which is made up of a set of parameters:

$$FV_{BLOB} = \left\{ \begin{array}{l} lb, (wm_1, wm_{\theta1}, wm_{\rho1}, ws_{\theta1}, ws_{\rho1}, fb_1, \alpha_1, e_{b1}), ..., \\ (wm_{lb}, wm_{\theta lb}, wm_{\rho lb}, ws_{\theta lb}, ws_{\rho lb}, fb_{lb}, \alpha_{lb}, e_{blb}) \end{array} \right\} \tag{20}$$

The feature vector for each character consists of two parts. The first contains a number of separate areas and subvectors storing the values of other parameters.

## 2.4   The Analysis of the Results

We analyzed the number of separate areas for the images from their own base. This allowed to determine number of areas for all character images from our database. Table 3 shows the number of compartments separated areas for particular classes of the image characters. Due to the large diversity it was decided to limit the number of areas to be taken into account while classification. Have been determined the number of features vector of extracted areas. In addition, based on the number of extracted areas realized preliminary classification, in the main classification process will be compared only feature vectors of the same length. The Fig. 5 shows examples of extracted areas of parametric representation for selected character images (Table 4).

---

[2] The coefficient ranges (0,1).

**Table 3.** The numbers of isolated areas

| Character | 0 | 1 | 2 | 3 | 4 | 5 | 6 | 7 | 8 | 9 |
|---|---|---|---|---|---|---|---|---|---|---|
| extracted areas $lb$ | 2÷7 | 1÷4 | 2÷6 | 2÷8 | 1÷7 | 3÷8 | 1÷8 | 1÷4 | 1÷8 | 2÷6 |

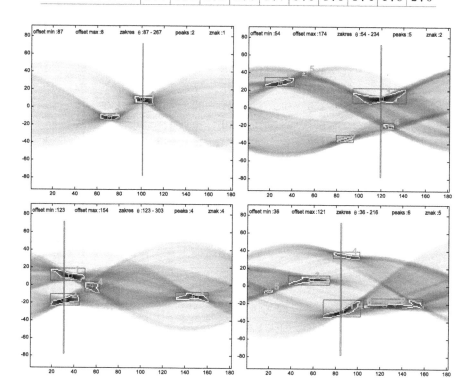

**Fig. 5.** The example of extracted areas from parametric representation

## 3    Summary

This work presents new approach to realise the support for automatic character recognition system, especially applied in the automatically recognition of post mail code. Although, this research area is well known and explored, with many effective examples of both scientific and commercial implementations, but automatic mail sorting systems are still insufficient.

In the vast majority of optical character recognition methods are based on modified quadratic discriminant function, hidden Markov models, normalized Fourier descriptors or MLP-SVM techniques. In our article, the idea and implementation of use of the parametric transformation in the process of zip code recognition for postal applications were presented.

The main advantages of the proposed in our article methods are: the finding geometric relations in the parameter space, invariance to background noise, low computational complexity, working with gray scale images of the zip code digits.

**Table 4.** The feature vectors $FV_{BLOB}$

| Parameter/Character | 0 | 1 | 2 | 3 | 4 | 5 | 6 | 7 | 8 | 9 |
|---|---|---|---|---|---|---|---|---|---|---|
| $l_{max}$ | 4 | 2 | 4 | 4 | 4 | 4 | 2 | 3 | 2 | 3 |
| $w_{m_1}$ | 1 | 1 | 1 | 1 | 1 | 1 | 1 | 1 | 1 | 1 |
| $w_{m\theta_1}$ | 121 | 101 | 120 | 56 | 30 | 85 | 132 | 130 | 65 | 59 |
| $w_{m\rho_1}$ | 68 | 83 | 81 | 97 | 109 | 121 | 108 | 87 | 82 | 88 |
| $w_{s\theta_1}$ | 109 | 102 | 119 | 44 | 32 | 85 | 137 | 130 | 76 | 56 |
| $w_{s\rho_1}$ | 65 | 83 | 79 | 101 | 109 | 118 | 108 | 87 | 77 | 88 |
| $fb_1$ | 1177 | 99 | 267 | 632 | 139 | 173 | 315 | 122 | 361 | 320 |
| $\alpha_1$ | 5 | −6 | 7 | 23 | 15 | 26 | 8 | −13 | 15 | −10 |
| $e_{b1}$ | 0,99 | 0,9 | 0,97 | 0,97 | 0,96 | 0,98 | 0,95 | 0,89 | 0,97 | 0,88 |
| $w_{m_2}$ | 0,89 | 0,89 | 0,73 | 0,98 | 0,99 | 0,76 | 0,53 | 0,84 | 0,85 | 0,66 |
| $w_{m\theta_2}$ | 110 | 71 | 29 | 110 | 36 | 58 | 96 | 77 | 147 | 138 |
| $w_{m\rho_2}$ | 134 | 105 | 64 | 52 | 82 | 84 | 94 | 81 | 100 | 61 |
| $w_{s\rho_1}$ | 86 | 70 | 28 | 110 | 35 | 57 | 94 | 78 | 132 | 132 |
| $w_{s\rho_2}$ | 136 | 103 | 63 | 51 | 80 | 84 | 95 | 78 | 102 | 60 |
| $fb_2$ | 502 | 81 | 115 | 166 | 165 | 127 | 39 | 68 | 546 | 140 |
| $\alpha_2$ | 2 | −2 | 16 | −14 | −19 | 4 | 0 | 18 | 12 | 2 |
| $e_{b2}$ | 0,99 | 0,94 | 0,97 | 0,98 | 0,97 | 0,98 | 0,97 | 0,94 | 0,96 | 0,94 |
| $w_{m_3}$ | 0,78 | | 0,6 | 0,78 | 0,93 | 0,76 | | 0,73 | | 0,65 |
| $w_{m\theta_3}$ | 35 | | 128 | 107 | 150 | 91 | | 91 | | 109 |
| $w_{m\rho_3}$ | 52 | | 112 | 131 | 104 | 58 | | 117 | | 142 |
| $w_{s\theta_3}$ | 54 | | 127 | 106 | 147 | 90 | | 92 | | 104 |
| $w_{s\rho_3}$ | 53 | | 112 | 132 | 104 | 57 | | 117 | | 141 |
| $fb_3$ | 554 | | 25 | 183 | 108 | 72 | | 67 | | 64 |
| $\alpha_3$ | −7 | | −10 | 3 | −7 | −12 | | −1 | | −11 |
| $e_{b3}$ | 0,99 | | 0,84 | 0,99 | 0,95 | 0,98 | | 0,94 | | 0,95 |
| $w_{m_4}$ | 0,52 | | 0,56 | 0,73 | 0,58 | 0,7 | | | | |
| $w_{m\theta_4}$ | 2 | | 87 | 130 | 53 | 124 | | | | |
| $w_{m\rho_4}$ | 137 | | 126 | 106 | 91 | 113 | | | | |
| $w_{s\theta_4}$ | 1 | | 86 | 145 | 54 | 127 | | | | |
| $w_{s\rho_4}$ | 137 | | 127 | 97 | 93 | 113 | | | | |
| $fb_4$ | 6 | | 9 | 386 | 62 | 155 | | | | |
| $\alpha_4$ | 90 | | 0 | 26 | −36 | 1 | | | | |
| $e_{b4}$ | 0,75 | | 0 | 0,99 | 0,88 | 0,99 | | | | |

While the disadvantages are: low value of the rejections, need to use some processing in initial stages of image processing. In further work the authors will include other blob features theory and upgraded to all alphanumeric signs to fully supported all data area from the postal items.

# References

1. Aissaoui, A., Haouari, A.: Normalised fourier coefficients for cursive arabic script recognition. Appl. Signal Process. **6**(3), 115–122 (1999)
2. Averbuch, A., Coifman, R., Donoho, D., Israeli, M., Walden, J.: Fast slant stack: A notion of radon transform for data in a cartesian grid which is rapidly computible, algebraically exact, geometrically faithful and invertible. SIAM Sci. Comput. (2001)
3. Haralock, R.M., Shapiro, L.G.: Computer and Robot Vision. Addison-Wesley Longman Publishing Co., Inc., Boston (1991)
4. Miciak, M.: Character recognition using radon transformation and principal component analysis in postal applications. In: 2008 International Multiconference on Computer Science and Information Technology, IMCSIT 2008, pp. 495–500. IEEE (2008)
5. Miciak, M.: Radon transformation and principal component analysis method applied in postal address recognition task. IJCSA **7**(3), 33–44 (2010)
6. Toft, P.A., Sørensen, J.A.: The Radon transform-theory and implementation. Ph.D. thesis, Technical University of DenmarkDanmarks Tekniske Universitet, Department of Informatics and Mathematical ModelingInstitut for Informatik og Matematisk Modellering (1996)
7. Venturas, S., Flaounas, I.: Study of radon transformation and application of its inverse to NMR. Algorithms in Molecular Biology 4 (2005)

# Lossless Compression Method for Digital Terrain Model of Seabed Shape

Wojciech Maleika and Paweł Forczmański[(✉)]

Faculty of Computer Science and Information Technology,
West Pomeranian University of Technology,
Żołnierska Street 49, 71–210 Szczecin, Poland
{wmaleika,pforczmanski}@wi.zut.edu.pl

**Abstract.** Dealing with Digital Terrain Models requires storing and processing of huge amounts of data, obtained from hydrographic measurements. Currently no dedicated methods for DTM data compression exist. In the paper a lossless compression method is proposed, tailored specifically for DTM data. The method involves discarding redundant data, performing differential coding, Variable Length Value coding, and finally compression using LZ77 or PPM algorithm. We present the results of experiments performed on real-world hydrographic data, which prove the validity of the proposed approach.

**Keywords:** DTM · Lossless compression · Differential coding · VLV Coding

## 1 Introduction

Contemporary hydrographic measurements increasingly often produce immense volumes of measurement data, which are postprocessed using specialised software. The measurement devices used during sea surveys, such as multibeam echosounders, perform readouts of millions of points during a single survey [9]. Due to the huge amount of acquired data we are able to create very accurate seabed models (Digital Terrain Models - DTMs) [15]. The processing of such data often involves the transformation of each point from the irregular grid to the regular one. In practice, one of several interpolation methods is used [11,18]. Obtained data are often used to create charts, cross-sections, volumetric analyses and planning offshore construction sites.

In most countries the regular surveys are performed in the areas of ship routes, in the neighbourhood of harbours, in canals, docks, moorings, navigable sections of rivers, etc. On the other hand, the measurements performed on an open sea are not common, which is caused by high financial cost and huge volume of data to process and store. But the future trend is certainly related to this field of exploration (especially when it comes to bays and smaller seas). Hydrographic work associated with sea floor modelling is a very sophisticated and long process which involves large technological cost. Developed models have to meet

R.S. Choraś (ed.), *Image Processing and Communications Challenges 8*,
Advances in Intelligent Systems and Computing 525, DOI 10.1007/978-3-319-47274-4_18

the highest standards [7] to ensure the safety of navigation. Thanks to advanced equipment (multibeam echosounders, precise positioning systems) and complex numerical algorithms, we get high quality output data. However, the constant increase in its volume (retaining the quality) makes a problem of compression very up-to-date. Hence, the proposed compression algorithm should meet specific requirements, such as high compression ratio and accurate reconstruction and make the measurements of larger areas possible.

The paper is organized as follows. First we introduce some problems of hydrographic data acquisition and processing, then we discuss possible ways of introducing general-purpose compression algorithms to the field od sea bed data compression. Finally, we propose a dedicated compression algorithm that makes use of characteristic features of seabed shape data. We finish the paper with some numerical experiments leading to the final conclusions.

## 2    Previous Works

The distinct characteristics of the seabed shape, slightly different measurement devices and varying requirements regarding the accuracy and the purpose of created models cause that the universal models and algorithms aimed at Digital Terrain Models can not always be applied for seabed shape. Moreover, the comparison of available algorithms for land and sea data is not always justified, since good results obtained for sea data might not be confirmed for land data, and opposite. Also the information representation within the model might be different – quite often within various solutions a grayscale map is used for determining the elevation map, whereas for description of the seabed using DTM practically always an array of real values with a given precision is utilised. Numerous papers can be found in the literature dealing with the compression of elevation data (mostly land), often gathered using the LIDAR system, where the processed elevation data are stored in grayscale images. The authors of [12] proposed the TIN DEM compression using second generation wavelets. In [16] the ODETLAP method was presented, applied to the compression and restoration of terrain data used for grid data. In [4] different methods of DEM images were described. In [19] a surface compression using overdetermined Laplacian approximation was presented. In [14] a method for DTM data reduction using self-organizing artificial neural networks was described. As it can be clearly seen, most of the works deal with compressing elevation data stored in images (bitmaps). There are only a couple of works describing lossless compression of terrain surfaces based on DTM or DEM, where not the images, but the raw data (real values) are compressed. In [5] the method of data reduction through spatial decompositions was described. In [1] DEM describing mountains was compressed using Huffman coding. The paper [2] investigates various lossy and lossless compression methods that can be applied to multibeam sonar data to reduce the size of acquired files without losing relevant information. In our previous works we developed a lossless compression method for measurement data (coming directly from MBES device) stored in ASCII files [10] and a near-lossless compression based on Principal Component Analysis [3]. That methods were based on similar assumptions and employed similar elementary processing stages. Obtained results are superior to the general-purpose compression methods, i.e. ZIP or RAR. On the other hand, the method described in this paper is oriented at different data type stored in different structure (gird).

In spite of many existing research dealing with lossy and lossless compression of DEM and DTM data, comparing the results presented in this paper with the results of other authors is practically impossible. The researchers utilize for the experiments their own test data, gathered using various devices, modeling diverse areas, and describing terrain using varying means and accuracy (using different data structures, including images, DTM, TIN, floating/integer values). Such a significant variety of test data makes it virtually impossible to properly compare the results. The problem lies also in the lack of common, freely accessible benchmark database of DEM and DTM data, which could serve as a reference dataset (as it is common with images databases). In order to evaluate the proposed method, the obtained results were compared to the results of compressing the same data using the commonly known lossless compressors such as ZIP (using Liv-Zempel [20] method and Huffmann coding) and RAR (based on Prediction by Partial Matching [13] algorithm).

## 3    Method Description

### 3.1    Assumptions

The most important criterion in seabed modelling is accuracy, expressed as errors: differences between each point of the DTM and the corresponding point on the real bottom. Bathymetric survey should allow the assessment of the model accuracy. The general error of the modelling is the result of errors made during the subsequent stages of DTM building, namely errors of the measurement device (depending on depth and type of seabed, model of the device – usually specified by the producer) [8], errors caused by survey parameters (speed, track configuration, echosounder parameters – difficult to assess, usually ignored), position errors (depend on positioning system), errors caused by the interpolation process [11] and smoothing (ignored so far). The necessity of error assessment for hydrographic works is caused by the requirements of high reliability of maps. Maximum acceptable error values are given by the International Hydrographic Organization. In practice, during the measurement of areas where under-keel clearance is critical, given the depth of approximately 10–20 m (typical for channels, port basins etc.), the total vertical uncertainty (TVU) equals approximately 20–30 cm. For deeper areas where under-keel clearance is not considered to be an issue it is much higher.

### 3.2    DTM's Data Files Characteristics

A DTM is often represented with a matrix of real numbers, stored with the precision of 4–12 fractional digits (depending on the software used). For each matrix element column and row indices determine the spatial location, and the value equals to the depth at a given point. The grid files can be saved either in ASCII, or binary formats. In the former case, each depth value is stored using 17 bytes. Additionally, each row of data ends with an end of line character. In case of a binary format, each depth value uses 8 bytes, there are no end of line markers. In both cases a short header must also be stored, containing some basic information regarding the file. The header's size is negligible and will be ignored during further analysis. Describing the depth using such high precision (12 fractional digits) seems to be excessive and pointless. Since the devices perform measurements with the accuracy of several centimetres, creating DTM with 1 cm precision is practically impossible (the RMS error when creating a DTM is often equal 5–10 cm) [8]. Such a detailed description of depth values within the files arises from the data structures used (floating point numbers) and is the result of certain precise calculations (e.g. the interpolation within the process of DTM building). We put a hypothesis, that the data gathered within the grid structure, describing the shape of seabed surface, may be adequately described with a 1 mm precision (i.e. 3 factional digits), and that such a truncation can be deemed lossless (i.e. not impacting the accuracy of created models). The proposed truncation can introduce (in case of special areas) the maximum models distortion of 0.005 % (which in turn equals to 0.3 % of the maximum TVU error allowed by the IHO recommendations). Such low values confirm, that the proposed storage changes can be considered lossless (i.e. they will have no real impact on the quality of the resulting model).

### 3.3   Preliminary Analysis of Algorithm Components

As a result of data truncation, according to the characteristics presented in above sections, we can obtain the compression efficiency (understood as a ratio of compressed file size to the original one) of 41 % for ASCII representation (24 bytes instead of 51, for each measurement). In case of binary format, storing 4, instead of 8 bytes, gives 50 % compression efficiency. The results of such a compression (reduction) for test data files are presented in the experimental part of the paper. It should be also expected, that a significant compression ratio can be achieved using known, general-purpose lossless compression methods, oriented at removing data redundancy, i.e. LZ77 [20] or PPM [13]. In order to verify this hypothesis, the test files were compressed using popular ZIP and RAR compressors. It occurred, that in case of typical sea bed data, that give an approximate efficiency of 17 %–22 %, for ASCII and 21 %–25 % for binary files, respectively. The statistical features of such files are also a root of further high compression thanks to the introduction of Huffmann coding. Hence, over five times decrease in size of data files may be considered satisfactory in many cases.

### 3.4   Proposed Differential Data Coding Using Variable Byte Length

Since the above-presented methods do not take into consideration the characteristics of processed data, we assumed that it is possible to achieve even higher compression efficiency by introducing a specific reorganisation of data file, including differential coding of depth values. Differential coding allows for a more effective utilisation of memory and a significant diminishing of file size, since the differences between depths may be stored using lower number of bytes. Here we propose a novel approach which includes the following operations:

1. Fixed point numbers of depth are converted to integer numbers by removing the decimal point (or more formally: by multiplying location values by 1000).
2. All the depths except the first one are stored as differences between current and previous values.
3. Subsequent measurements are additionally encoded. Since most of the values are small, it could be beneficial to use a technique of storing the numbers using variables of variable length. Obviously, the information about the actual number of occupied bytes must be added. For this purpose a modified Variable Length Value Coding method (VLV) [6] was utilised. In this case each VLV value is stored on byte-wide words, containing two portions: 7 bits of the actual information and one bit denoting possible continuation. If the most significant bit (continuation bit) is set, then the number is continued in the next byte. Otherwise, this is the last byte of a number. In order to encode a number in VLV, it needs to be divided into 7-bit long groups; then each group is appended with the continuation bit. In such case all the numbers within the range $< -63; 64 >$ are stored using one byte, numbers within the range $< -4096; 4096 >$ using two bytes, and so on. In order to retrieve a number encoded using VLV, the continuation bit must be removed, remaining bits must be concatenated to the number being formed, until the final byte is encountered.

The decompression procedure is very similar. At the first stage subsequent bytes are read from a file and decoded, then they are converted from differential form into plain values (actual depth values).

As a result of a single cycle of coding/decoding operations, we obtain identical file, hence we can speak about lossless data compression. The data obtained after conversion to differential form and stored using VLV algorithm can be further processed by LZ77 (ZIP) or PPM (RAR) compression, in order to minimize their redundancy. The effectiveness evaluation of the proposed algorithm is presented further in the paper.

## 4    Experiments

The compression experiments were performed on several datasets created using the measurements collected from Szczecin Lagoon and Pomeranian Bay (courtesy of Maritime Office in Szczecin, Poland). The parameters of created surfaces (grids) are presented in Table 1, while their 3D visualizations – in Fig. 1. As it can be seen, each surface has different geomorphological properties, for example "Gate" is a visualization of a route gate, "Wrecks" presents an area with car wrecks, "Swinging" is a place where ships can rotate and "Anchorage" is the flat anchorage ground. They were intentionally chosen to cover mostly encountered types of sea floor. While the sea floor in most areas in the world is not so variable, the results of the experiments should be representative and give a good approximation of projected efficiency of the method. It can be easily noticed, that the data gathered within a grid structure can be significantly compressed. Regardless of the selected storage format (binary or ASCII) the application of commonly known lossless compressors leads to the reduction of file size to approximately 17–25 % of the original one. Such an approach is currently widely used, especially for backup purposes or for transferring of larger amounts of data over the network. The file structure analysis and the inspection of the DTM models' properties allowed us to develop a novel DTM compression method, utilising the algorithm of redundant data reduction, differential encoding and VLV coding. Incorporating those stages into the data storage process can lead to the further increase in the compression efficiency.

**Table 1.** Characteristics of benchmark surfaces

| Surface | Grid | | | Filesize [MB] | |
|---|---|---|---|---|---|
| | Resolution [m] | Area [m] | No. points [thousands] | ASCII | Binary |
| Achorage | 1 | 3008 × 1696 | 5101 | 82.71 | 38.92 |
| Swinging | 0.75 | 2464 × 1760 | 4336 | 70.31 | 33.09 |
| Gate | 0.5 | 1888 × 1632 | 3081 | 49.95 | 23.51 |
| Wrecks | 0.01 | 1856 × 672 | 1247 | 20.22 | 9.52 |

Finally, such processed data can be further compressed using general-purpose algorithms, which leads to the final compression ratio of approximately 3–7 %. Based on the obtained results it can be noticed, that the surfaces with a smooth changes structure, such as "Anchorage", can be compressed to a smaller size, reaching the size of 2.9 % of original ASCII file (Diff+VLV+RAR). For the surfaces with some more rapid changes, such as "Wrecks", the compression efficiency equals 4 % (Diff+VLV+RAR). For a highly diverse surfaces, with numerous significant changes of depths, such as "Swinging" and "Gate", the compression efficiency is smallest and falls within 5–6 % interval. Given that the amount of data gathered within a grid structure is significant, and the fact, that those data are expressed using real numbers, the almost tenfold reduction of data size achieved by applying the proposed compression method, dedicated for this specific kind of data, should be considered a satisfactory result. The variance of compression efficiency is very small in case of plain Diff+VLV approach, while it increases in case of additional ZIP/RAR compression. An additional analysis of the obtained results shows, that approximately 80–95 % of depth values are stored using only one byte, 5–19 % using 2 bytes, under 1 % using 3 bytes and under 0.05 % using 4 or more bytes.

The comparison with the results obtained using ZIP or RAR method shows, that the compression ratio when using differential and VLV coding is higher. This

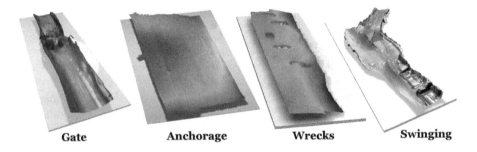

| Gate | Anchorage | Wrecks | Swinging |

**Fig. 1.** Benchmark surfaces used in the experiments

**Table 2.** Compression efficiency calculated as a percentage of compressed filesize to the original one for the developed method compared with other approaches

| Surface | ASCII | | | Binary | | | | Diff+VLV | | |
|---|---|---|---|---|---|---|---|---|---|---|
| | Trunc. | +ZIP | +RAR | Orig. | Trunc. | +ZIP | +RAR | Plain | +ZIP | +RAR |
| Anchorage | 41 | 21.10 | 16.92 | 47.07 | 23.54 | 11.15 | 9.71 | 13.89 | 4.10 | 2.94 |
| Swinging | 41 | 21.75 | 17.58 | 47.06 | 23.53 | 12.39 | 10.33 | 15.25 | 6.96 | 5.89 |
| Gate | 41 | 22.97 | 17.35 | 47.06 | 23.54 | 12.84 | 10.78 | 14.68 | 6.07 | 5.50 |
| Wrecks | 41 | 22.65 | 17.06 | 47.08 | 23.54 | 12.71 | 10.58 | 14.34 | 4.45 | 4.01 |
| Mean | 0 | 22.12 | 17.23 | 47.07 | 23.54 | 12.27 | 10.35 | 14.54 | 5.40 | 4.59 |
| Variance | 0 | 0.73 | 0.09 | 0 | 0 | 0.60 | 0.22 | 0.32 | 1.83 | 1.86 |

leads to a conclusion, that the method based on differential coding combined with coding using variable number of bytes is well adjusted to the characteristics of the sea survey measurement data (Table 2).

## 5   Conclusions

The method proposed in this paper consists of initial reorganisation of data file by discarding redundant information, followed by calculating differences between measurement depths and encoding them using variable number of bytes, and finally compressing using LZ77 or PPM algorithm. Such a procedure leads to significantly improved compression results. The compression efficiency reaches about 10 %, which means reduction by order of magnitude. In practise the whole procedure is reasonably fast, overall processing time is shorter than for ZIP compression alone. The developed algorithm may be used in hydrographic software as additional functionality for saving DTM's data. Its utilisation may significantly reduce the amount of stored data and speed up data transfer in computer networks, while still maintaining acceptable compression time.

The future works may include the application of other methods of lossless compression, adapted to this particular problem, i.e. adaptive predictors [17].

# References

1. Chang, Z.Q., Wu, L.X.: Mountain grid DEM data compression based on wavelet transform and mixed entropy coding. Geogr. Geo-Inf. Sci. **1**, 24–27 (2004)
2. Chybicki, A., Moszynski, M.: Applications of compression techniques for reducing the size of multibeam sonar records. In: Proceedings of the 2008 1st International Conference on Information Technology, pp. 1–4 (2008)
3. Forczmański, P., Maleika, W.: Near-lossless PCA-based compression of seabed surface with prediction. In: Kamel, M., Campilho, A. (eds.) ICIAR 2015. LNCS, vol. 9164, pp. 119–128. Springer, Heidelberg (2015). doi:10.1007/978-3-319-20801-5_13
4. Franklin, W.R., Said, A.: Lossy compression of elevation data. In: Seventh International Symposium on Spatial Data Handling, Delft (1996)
5. Gerstner, T.: Multiresolution compression and visualization of global topographic data. GeoInformatica **7**(1), 7–32 (2003)
6. Gordon, V., Cormack, R., Horspool, R.N.: Algorithms for adaptive huffman codes. Inf. Process. Lett. **18**(3), 159–165 (1984)
7. International Hydrographic Organization standards for hydrographic surveys. Pub. No. 44, 5th Ed. (2008). http://www.iho.int//iho_pubs//standard//S-44_5E.pdf
8. Maleika, W.: Development of a method for the estimation of multibeam echosounder measurement accuracy. Przegląd Elektrotechniczny **88**(10b), 205 (2012)
9. Maleika, W., Czapiewski, P.: Visualisation of multibeam echosounder measurement data. In: Maji, P., Ghosh, A., Murty, M.N., Ghosh, K., Pal, S.K. (eds.) PReMI 2013. LNCS, vol. 8251, pp. 373–380. Springer, Heidelberg (2013). doi:10.1007/978-3-642-45062-4_51
10. Maleika, W., Czapiewski, P.: Lossless compression method for ASCII UTM format sea survey data obtained from multibeam echosounder. Roczniki Geomatyki, Vol. XII, 3(65), 289–301 (2014)
11. Maleika, W.: Moving average optimization in digital terrain model generation based on test multibeam echosounder data. Geo Mar. Lett. **35**(1), 61–68 (2015)
12. Pradhan, B., Mansor, S.: Three dimensional terrain data compression using second generation wavelets. In: 8th International Conference on Data, Text and Web Mining and Their Business Applications. Wit Transactions on Information and Communication Technologies, 38 (2007)
13. Shkarin, D.: PPM: one step to practicality. In: Proceedings of the Data Compression Conference, pp. 202–211 (2002)
14. Stateczny, A., Wlodarczyk-Sielicka, M.: Self-organizing artificial neural networks into hydrographic big data reduction process. In: Kryszkiewicz, M., Cornelis, C., Ciucci, D., Medina-Moreno, J., Motoda, H., Raś, Z.W. (eds.) RSEISP 2014. LNCS (LNAI), vol. 8537, pp. 335–342. Springer, Heidelberg (2014). doi:10.1007/978-3-319-08729-0_34
15. Stephens, D., Diesing, M.: A comparison of supervised classification methods for the prediction of substrate type using multibeam acoustic and legacy grain-size data. PloS ONE **9**(4), e93950 (2014)
16. Stookey, J., Xie, Z., Cutler, B., Franklin, W., Tracy, D., Andrade, M.: Parallel ODETLAP for terrain compression and reconstruction. In: GIS 2008: Proceedings of the 16th ACM SIGSPATIAL International Conference on Advances in Geographic Information Systems (2008)
17. Ulacha, G., Stasinski, R.: New context-based adaptive linear prediction algorithm for lossless image coding. In: International Conference on Signals and Electronic Systems (2014)

18. Wawrzyniak, N., Hyla, T.: Managing depth information uncertainty in inland mobile navigation systems. In: Kryszkiewicz, M., Cornelis, C., Ciucci, D., Medina-Moreno, J., Motoda, H., Raś, Z.W. (eds.) RSEISP 2014. LNCS (LNAI), vol. 8537, pp. 343–350. Springer, Heidelberg (2014). doi:10.1007/978-3-319-08729-0_35

19. Xie, Z., Franklin, W., Cutler, B., Andrade, M., Inanc, M., Tracy, D.: Surface compression using over-determined laplacian approximation. In: Proceedings of the SPIE, Advanced Signal Processing Algorithms, Architectures, and Implementations XVII, vol. 6697 (2007)

20. Ziv, J., Lempel, A.: Universal algorithm for sequential data compression. IEEE Trans. Inf. Theory **23**(3), 337–343 (1977)

# On Combining Dual Morphological Binary Operators Using Median Set

Marcin Iwanowski[✉]

Institute of Control and Industrial Electronics, Warsaw University of Technology,
ul. Koszykowa 75, 00-662 Warszawa, Poland
marcin.iwanowski@ee.pw.edu.pl

**Abstract.** In the paper an approach to combining dual morphological operators: erosion/dilation and opening/closing is discussed. It is based on the morphological interpolation by means of a median set. The boundary of such a set is located halfway between the boundaries of sets that are results of dual operators. The proposed combination of dual morphological operators by median set is an self-dual morphological operator able to process both foreground and background image details in the same way. An application of the discussed approach to binary image enlargement is presented in the paper as well.

**Keywords:** Morphological image processing · Interpolation · Duality · Self-dual operators

## 1 Introduction

Morphological operators [7,8,11] are defined pairwise in such a way that an extensive operator is accompanied by its anti-extensive counterpart. When applied to binary image filtering, single operator influences only some image regions e.g. foreground. In order to get some influence on the background, the dual operator should be applied. In order to influence both, foreground and background image regions, extensive and anti-extensive operators are combined together. The typical way of combining both operators is cascade-wise e.g. the given operator is followed by its dual version [8].

In this paper a novel way of combining the results of mutually dual operators is discussed, that makes use of morphological interpolation. In particular, morphological median operator is considered, which allows obtaining the intermediary image between results of both operators. In case of binary images it results in an intermediary shape, the boundary of which is equally distant from boundaries of results of both operators: extensive and anti-extensive.

Morphological approach to image interpolation was formulated in 90's of XX century. There are two principal area of exploration in this domain: inter-frame – interpolating between two images and intra-frame – interpolating to fill empty regions within single image. The interframe interpolation methods consist in creation of the intermediary two-dimensional images between two given ones.

© Springer International Publishing AG 2017
R.S. Choraś (ed.), *Image Processing and Communications Challenges 8*,
Advances in Intelligent Systems and Computing 525, DOI 10.1007/978-3-319-47274-4_19

Two principal approaches have been given. The first one, based on the morphological median set, was presented in [1,6]. It may be applied to any kind of image: binary, mosaic, graytone and color [2]. Another approach is represented by the interpolation function method introduced in [4]. This method is based on the function which describes the relative distance between the objects and can be applied to binary and mosaic images. In the former case the morphological interpolation can be combined with the affine transform [3]. A different approach to morphological interpolation was presented in e.g. [10] – an 'intraframe' interpolation. It deals with a single incomplete image and it reconstructs the image surface starting from the contour lines. This approach makes use of the geodesic distance function obtained by the geodesic propagation. The proposed application of morphological interpolation combines both meanings of interpolation – it makes use of the median set within a single image.

The proposed way of combining the dual operators may be used to smooth the contours of image objects for e.g. shape simplification. It can also be used for enlarging binary shapes. In the latter case, the shape rescaled using nearest neighbor approach (that may have blocky appearance) is further smoothed by the proposed approach.

The paper is organized as follows. In Sect. 2 the background notions related to duality, dual morphological operators and median set are recalled. In the Sect. 3, the idea of combining the dual operators by means of median set is presented along with some application examples. Section 4 concludes the paper.

## 2    Background Notions and Definitions

The research, results of which, are presented in this paper are focused on binary images. Formally, such an image will be defined as as two sets of pixels: foreground, and its complement – background. In the case of an image function $f : D \rightarrow \{0,1\}$ ($D$ is the image definition domain) both sets are defined as, respectively: $F = \{p : f(p) = 1\}$, $\overline{F} = \{p : f(p) = 0\}$. The following properties are obvious: $F \cap \overline{F} = \emptyset$ and $F \cup \overline{F} = D$.

### 2.1    Duality of Morphological Operators

The morphological image operators are based on local minimum and maximum computations within pixel neighborhood defined by the structuring element. Owing to that fact, they are defined in pairs of dual operators. The duality is defined in such a case in the following way:

$$\overline{F} \triangle B = \overline{F \bigtriangledown B}, \tag{1}$$

where $F$ stands for an binary image, $\overline{F}$ for its complement and $\triangle, \bigtriangledown$ for a pair of dual operators. Such operators are usually characterized by the property of being extensive and anti-extensive. These properties are defined as follows (extensive and anti-extensive operators, respectively):

$$F \subseteq F \triangle B \; ; \; F \bigtriangledown B \subseteq F. \tag{2}$$

The duality of pairs of increasing/decreasing morphological operators is considered often as their disadvantage. In particular, it plays an important role if the content of an image is to be treated without differentiating darker and lighter image regions. In order to treat both cases symmetrically the image processing operator $\triangle$ should be self-dual, which means that:

$$\overline{X \triangle B} = \overline{X} \triangle \overline{B}. \tag{3}$$

Contrary to morphological ones, most of the popular neighborhood operators, like linear or median filters are self-dual. The typical approach to combining dual morphological operators is an alternating or alternating-sequential filter that consist of extensive operator followed by its anti-extensive counterpart or vice-versa [5,8]. The problem of self-duality has also been studies within the mathematical morphology community, some interesting approaches has been proposed [9].

## 2.2    Pairs of Dual Morphological Operators

The basic morphological operators are erosion and dilation. Dilation refer to the local maximum operator that computes maximal value among pixels belonging to the neighborhood defined by the structuring element. In the case of binary images dilation is defined as:

$$F \oplus B = \bigcup_{b \in B} F_{[-b]}, \tag{4}$$

where $F_{[-b]}$ refers to shifting image $F$ by a vector $-b$. Dilation result in extending the binary image being its argument.

An operator dual to dilation is an erosion that, in turn, is the local minimum operator defined in the binary case as:

$$F \ominus B = \bigcap_{b \in B} F_{[-b]}, \tag{5}$$

which means that it results in shrinking of the input binary image.

Erosion and dilation are used to define most popular morphological filters of opening and closing that are defined as, respectively:

$$F \circ B = (F \ominus B) \oplus B^T; \; F \bullet B = (F \oplus B) \ominus B^T, \tag{6}$$

where $B^T = \{p : -p \in B\}$ stands for the transposed structuring element $B$. Opening and closing is the second pair of dual morphological operators, that are fulfilling the conditions of morphological filters (idempotence and order-preservation). Opening allows removing foreground objects from an binary image, while closing – background ones.

## 2.3   Median Set

Morphological median set has been introduced in [1] as a tool for interpolation between sets (binary images). It aims at providing the intermediary set, the boundary of which is located midway between boundaries of two input objects. Let $X$ and $Y$ be ordered pair of binary images ($X \subseteq Y$). Every point of the difference $Y \setminus X$ (point belonging to $Y$ but not to $X$) is characterized by two distances: to $X$ and to $\overline{Y}$. The first distance may be obtained by succesive by successive dilations of $X$, while the second by successive erosions of $Y$. Therefore, point $m$ at distance $\leq \lambda$ to $X$ and $\geq \lambda$ to $\overline{Y}$ belongs to $\{(X \oplus \lambda B) \cap (Y \ominus \lambda B)\}$. Hence $m$ belongs to the following binary image:

$$M = \cup_{\lambda \geq 0}\{(X \oplus \lambda B) \cap (Y \ominus \lambda B)\}; \ \lambda B = B \underbrace{\oplus B... \oplus B}_{\lambda - times} \tag{7}$$

This equation defines an image of morphological median set. It consists of all those points of $Y$, which are closer to $X$ than to $\overline{Y}$. In that case this notion is equivalent to the notion of the *influence zone* of $X$ in $Y$ defined by:

$$IZ_Y(X) = \{p : d(p, X) < d(p, \overline{Y})\} \tag{8}$$

Where $d$ is a distance from a point to the boundary of the object.

The notion of median set can be expanded to a more general case of two objects $X$ and $Y$ with a non-empty intersection ($X \cap Y \neq \emptyset$). If this condition is satisfied, the median set of two sets $X$ and $Y$ is defined as the influence zone of $(X \cap Y)$ in $(X \cup Y)$:

$$M(X, Y) = IZ_{(X \cup Y)}(X \cap Y) \tag{9}$$

Where influence zone $IZ$ is defined by the (8). Definition (7) has, in this case, the following form [6]:

$$M(X, Y) = \cup_{\lambda \geq 0}((X \cap Y) \oplus \lambda B) \cap ((X \cup Y) \ominus \lambda B) \tag{10}$$

It defines the median object of two input objects with a non-empty intersection.

Based on the definition expressed by Eq. 10, computation of median set can be performed using the iterative algorithm [1,2]. The initial values (iteration $i = 0$) of temporary images $Z_0, W_0, M_0$ are set-up as ($X$ and $Y$ are the input objects such that $X \cap Y \neq \emptyset$):

$$Z_0 = X \cap Y; \ W_0 = X \cup Y; \ M_0 = X \cap Y \tag{11}$$

New values in the $i$-th iteration - $Z_i$, $W_i$ and $M_i$ are calculated as:

$$Z_i = Z_{i-1} \oplus B; \ W_i = W_{i-1} \ominus B; \ M_i = (Z_i \cap W_i) \cup M_{i-1} \tag{12}$$

If $M_i \neq M_{i-1}$ – increment $i := i + 1$ and recalculate the images according to (12). Otherwise, if $M_i = M_{i-1}$, iterations stop and finally: $M(X, Y) = M_i$, median set is reached.

There exist an alternative way to produce a median set, based on distance functions. Two examples of median set are shown in Fig. 1.

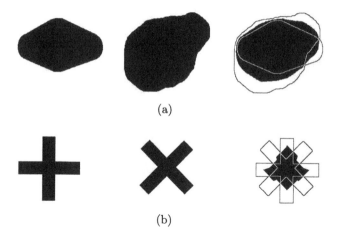

(a)

(b)

**Fig. 1.** Examples of median sets of almost-convex (a) and non-convex (b) input sets. Left and middle – input sets, right – median set (with outlines of input sets superimposed)

## 3    Combining Dual Operators Using Morphological Median

The morphological median has several valuable properties e.g. it is self-dual: $\overline{M(X,Y)} = M(\overline{X}, \overline{Y})$, symmetric $M(X,Y) = M(Y,X)$ and satisfies the following relation: $X \cap Y \subseteq M(X,Y) \subseteq X \cup Y$. Thanks to them, it may be used to combine a pair of dual morphological operators. We got in such a way the following operators that processes the original binary image $X$:

$$M_B^{(\ominus\oplus)}(X) = M(X \oplus B, Y \ominus B), \tag{13}$$

$$M_B^{(\circ\bullet)}(X) = M(X \circ B, X \bullet B), \tag{14}$$

where $B$ stand for the structuring element used in both dual operators. The properties of the proposed operators are discussed below.

– *Self-duality* of the above operators can be easily proven using the property of duality of a pair of morphological operators and self-duality of the median set:

$$M_B^{(\ominus\oplus)}(\overline{X}) = M(\overline{X} \oplus B, \overline{X} \ominus B) = M(\overline{X \ominus B}, \overline{X \oplus B}) \tag{15}$$

$$= \overline{M(X \ominus B, X \oplus B)} = \overline{M_B^{(\ominus\oplus)}(X)} \tag{16}$$

Similar proof may be constructed for the opening/closing case, which shows that:

$$M_B^{(\circ\bullet)}(\overline{X}) = \overline{M_B^{(\circ\bullet)}(X)}, \tag{17}$$

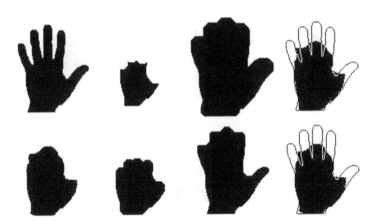

**Fig. 2.** An example results. On the top – original image, its erosion, dilation and median set of both (with the contour of original image superimposed). Bottom – opening/closing median set, result of opening, closing, opening/closing median set superimposed on the external contour of the original image

– The operators, contrary to four basic morphological ones are *neither extensive nor antiextensive*[1], which means that:

$$M_B(X) \notin X \; ; \; X \notin M_B(X). \tag{18}$$

– The operators, contrary to opening and closing (but not to erosion/dilation) are in general (apart from simple cases) *not idempotent*:

$$M_B(M_B(X)) \neq M_B(X). \tag{19}$$

An example of both operators applied to the same image is shown if Fig. 2. Results are similar in both cases.

The proposed approach may be used to remove from a binary image small objects, in the same turn, belonging to both foreground and background. It also results in some smoothing of boundaries of objects on the image. Thanks to this property it may be successfully used to smooth boundaries after binary image enlargement using nearest neighbor interpolation. An example of such operation is shown in Figs. 3 and 4. The input image is the result of enlarging the original one by 10 times. The nearest-neighbor enlarged image consist thus of $10 \times 10$ pixel blocks. An application of the median set of dual operators results in smooth boundaries. The smoothness level depends creates a 'gap' around the boundary. Within this gap, halfway between boundaries of dual operators, the boundary of a median set has been produced. The size of such a gap depends on the type of operation in the case of $M^{(\ominus\oplus)}$ the gap is wider, while in the $M^{(\circ\bullet)}$ it is narrower. The latter is due to the fact that opening and closing operators consists each of two dual base ones (erosion and dilation). Application of the second of the latter operators result in the gap shrinking.

---

[1] Erosion and opening are anti-extensive while dilation and closing are extensive.

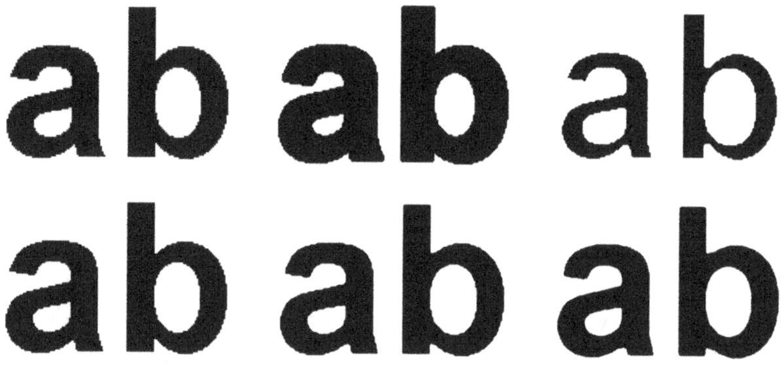

# ab ab ab
# ab ab ab

**Fig. 3.** An application of erosion-dilation median set to binary image enlargement. On the top: left – original image; -middle – dilation; - right – erosion. Bottom row – medians with the size of $B$ equal to 10 (left), 20 (middle), 30 (right)

**Fig. 4.** Enlarged parts of images from Fig. 3 – original image and median sets with the size of $B$ equal to 10 (left), 20 (middle), 30 (right)

## 4  Conclusions

In the paper, a novel approach to combining dual morphological operators was proposed. It is based on the median set, i.e. binary image the boundary of which is located halfway between boundaries of two input sets. As two input sets, the results of a pair of dual morphological operators are used. The result of such an operator is – contrary to the usual morphological operators – self-dual. In the paper the properties of the proposed operator was investigated, as well as some examples was provided.

The experiments carried-out show that the proposed approach allows not only for removal of image objects both foreground and background but also for smoothing the boundaries of input binary image. The latter property makes the method useful for correcting borders of object on binary images enlarged using the nearest-neighbor method.

# References

1. Beucher, S.: Interpolation of sets, of partitions and of functions. In: Heijmans, H., Roerdink, J. (eds.) Mathematical Morphology and its Application to Image and Signal Processing. Kluwer Academic Publishers (1998)
2. Iwanowski, M., Serra, J.: Morphological interpolation and color images. In: Proceedings of 10th International Conference on Image Processing, ICIAP 1999, 27–29 September 1999, Venice, Italy, pp. 50–55. IEEE, September 1999
3. Iwanowski, M., Serra, J.: The morphological-affine object deformation. In: Goutsias, J., Vincent, L., Bloomberg, D. (eds.) Mathematical Morphology and its Applications to Signal and Image Processing, pp. 81–90. Kluwer Academic Publishers, Boston (2000)
4. Meyer, F.: A morphological interpolation method for mosaic images. In: Maragos, P., Schafer, R., Butt, M. (eds.) Mathematical Morphology and its Applications to Image and Signal Processing, pp. 337–344. Kluwer Academic Publishers, Boston (1996)
5. Salembier, P., Garrido, L., Garcia, D.: Auto-dual connected operators based on iterative merging algorithms. In: Proceedings of the International Symposium on Mathematical Morphology and its Applications to Image and Signal Processing, pp. 183–190 (1998)
6. Serra, J.: Hausdorff distance and interpolations. In: Heijmans, H., Roerdink, J. (eds.) Proceedings of the Fourth International Symposium on Mathematical Morphology and its Applications to Image and Signal Processing, pp. 107–114. Kluwer Academic Publishers (1998)
7. Serra, J.: Image Analysis and Mathematical Morphology, vol. 1. Academic Press, London (1982)
8. Serra, J. (ed.): Image Analysis and Mathematical Morphology. II: Theoretical Advances, vol. 2. Academic Press, London (1988)
9. Soille, P.: Beyond self-duality in morphological image analysis. Image Vis. Comput. **23**(2), 249–257 (2005)
10. Soille, P.: Generalized geodesic distances applied to interpolation and shape description. In: Serra, J., Soille, P. (eds.) Mathematical Morphology and its Applications to Image and Signal Processing, pp. 193–200. Kluwer Academic Publishers, Netherlands (1994)
11. Soille, P.: Morphological Image Analysis: Principles and Applications, 2nd edn. Springer, Berlin (2003)

# Subjective Image Quality Assessment Optimization

Anna Lewandowska (Tomaszewska)[✉]

Faculty of Computer Science and Information Technology,
West Pomeranian University of Technology,
Zolnierska 49, 71-210 Szczecin, Poland
atomaszewska@wi.zut.edu.pl

**Abstract.** As the results of computer algorithms methods are often visual, image quality assessment is one of its central problems. To provide a convincing proof that a new method is better than the state-of-the-art the image quality assessment should be employed. Therefore image based projects are often accompanied by user studies, in which a group of observers rank or rate results of several algorithms. Unfortunately the problem posed by subjective experiments is their time-consuming and expensive nature. This paper is intended to present how to make the subjective experiments less expensive and therefore more usable.

**Keywords:** Image quality assessment · User-study · Scene understanding · Computer vision

## 1 Introduction

Image quality assessment is crucial in many graphics or image processing algorithms resulting with visual results. There is often a need to compare them with the state-of-the-art methods. Comparison via several examples included in the paper and carefully inspected with the results of competitive algorithms is an effective method, but only if the visual difference is unquestionably large. If the differences are subtle, such informal comparison is often disputable. The most reliable way of assessing the quality of an image is by subjective evaluation.

Given that the ultimate receivers of images are human eyes, the human subjective opinion is the most reliable value for indicating the image perceptual quality. For these reasons there is a strong new trend to support visual results by user studies, in which a larger group of assessors make their judgments about the preference of one method over another. There are numerous methods of subjective quality assessment, but all of them are time-consuming and expensive, which makes them impractical for most image applications. However, the obtained subjective rating value can be recognized as the ground truth of the image perceptual quality.

The goal of the paper, that complements our approach presented in [4], is indicating the subjective experiments efficiency by Pearson correlation between

© Springer International Publishing AG 2017
R.S. Choraś (ed.), *Image Processing and Communications Challenges 8*,
Advances in Intelligent Systems and Computing 525, DOI 10.1007/978-3-319-47274-4_20

full and reduced dataset of images. Therefore, the objective metrics, known as $FSIM_c$ [19] was employed to estimate the database redundancy. To check practical usage of the approach we verified a method on a standard well-known databases: LIVE [13], IVC_SubQualityDB [15], TID2013 [12] and CISQ [3].

The paper is organized as follows. In Sect. 2, previous works are discussed. The efficient method for subjective experiments and its improvement is presented in Sect. 3. Analysis of databases redundancy and usability of the approach tested on the well-known databases are presented in Sect. 4. The last section presents conclusions and suggestions for possible future work.

## 2   Previous Work

The subjective quality assessment are used practically in different applications [4,6,7,16–18]. However the problem posed by subjective experiments is their time consuming and expensive nature. Time compensation through the reduction of trial or scene numbers may be a solution to the problem. The number of trials can be limited using balanced incomplete block designs [1] in which all possible paired comparisons are indirectly inferred. But even more effective reduction of trials can be achieved if a sorting algorithm is used to choose pairs to compare [14]. As reported in [5,6], when the reduced design with sorting algorithm is used for the forced choice technique, the method is even faster than the single stimulus. However, the number of images selected for the experiment still influences the reduced number of trials. If the number is high, the experiments remain time-consuming. Therefore, the scene pre-selection is not an uncommon practice, and was proposed to measure quality scales (quality rulers) for the ISO 20462 standard [2]. Nonetheless, the methodology of this process is not well explained and analyzed.

The reduction of the number of required sessions per observer, by a semi-manual selection process was proposed in [8]. The use of adapted objective quality metrics to estimate the quality of data before the test may be a promising alternative. Such an approach was proposed in [9,10], where the problem of decreasing the number of video sequences was analyzed. Naturally, this raises questions about the reliability of the measure. The scene preselection was examined also in [4], where a method of limiting the number of scenes that need to be tested, by employing a clustering technique and evaluated it on the basis of compactness and separation criteria is proposed. The problem is that before starting the procedure of the scene reduction we do not have knowledge about database redundancy. In the paper the objective metrics MS-SSIM was used. However the MS-SSIM metrics have much lower correlation values for the most relevant datasets. In the paper the assumption that the metric should characterized by the Pearson's correlations about 0.9 was made. Therefore we used the $FSIM_c$ (Feature Similarity Index Measure) [19] method, as it is reported with high Pearson's correlations between MOS (*Mean Opinion Score*) and the objective scores.

# 3   The Most Prominent Experimental Method

The method of pairwise comparisons tends to be the most often used in graphics.

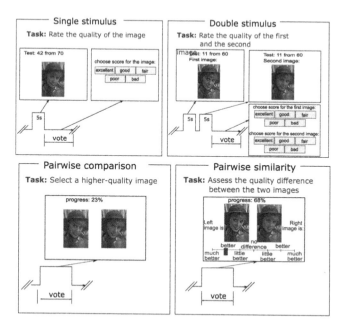

**Fig. 1.** Overview of the forced choice subjective quality assessment method. The diagram shows the timeline of the method and the corresponding screen

Intuitively, it could be expected that pairwise comparison method, where the observer has to choose a better of two images, is easier for observers, easier to reproduce, and thus more accurate than direct rating. This intuition was confirmed in a formal study [7], where four subjective methods of quality assessment were compared: single and double stimulus methods, forced choice pairwise comparison and similarity judgements. All these methods are visually illustrated in Fig. 1. In case of both the force-choice and similarity judgements methods, the number of comparisons was reduced using an efficient sorting algorithm [14]. Refer to the paper [7] for more details on the experimental setup.

The sensitivity of each method was measured in terms of the effect size $d$, which is defined as the difference between a pair of quality scores normalized by a common standard deviation. The values of effect size for each method are shown in Fig. 2. The forced choice pair-wise comparison method results in statistically significantly higher effect size as compared to both single and double stimulus methods. But the difference is not very large in practical terms; only in about 64 % of cases the forced choice method will result higher sensitivity than the double stimulus method, assuming the same sample size. Therefore, our

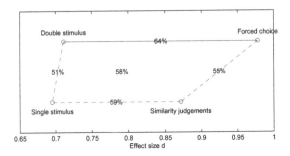

**Fig. 2.** The comparison of effect size for each experimental method. The larger the effect size, the more accurate the method is. The y-axis is used only to better layout the methods and show their relations. The percentages indicate the probability that for a random pair of scenes and distortion types, the method on the right will result in higher sensitivity than the method on the left. If the line connecting two conditions is red and dashed, it indicates that there is no statistical difference between this pair of conditions (Ho could not be rejected for $\alpha = 0.05$ and adjusted for multiple comparisons)

data shows that the difference between direct rating and pairwise comparison methods exists, but it does not seem to be as dramatic as the four-fold reduction of standard deviation reported in [11].

Even though the pair-wise comparison methods are marginally more sensitive, they have also the reputation of being tedious and requiring a very large number of trials. However, we found that when a reduced pairwise comparison design is used [14], pairwise comparison method can be significantly faster than direct rating methods, even for a large number of compared conditions [7].

## 4   Scene Pre-selection Method

The experimental results concerning many cases and conducted in different sessions are often noisy and their proper analysis and interpretation is not trivial. Instead of studying a large number of scenes, focusing the measurement on a few scenes that differ the most in their quality scores is proposed. The rationale is that measuring two scenes that result in very similar quality scores does not contribute to better understanding of how image content affects quality. Our main intention is to reduce the effort in algorithm evaluation by selecting only the representative scenes and running the full experiment on them. But to decide on the representative scenes, their quality should be known, which makes it necessary to run the experiment on all images. The image selection, however, can be relatively efficient if the decision can be based on a small sample collected for all images in a pilot experiment. By the pilot experiment, the first phase of experiment to choose representative images is understood. In the practical application, the subjective results to make the Pilot stage are unknown. Therefore to verify the procedure, the Pilot stage was computed parallel with two different data: subjective results and objective metrics.

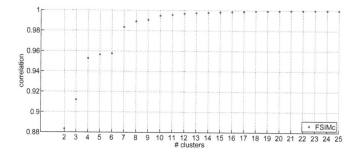

**Fig. 3.** Correlation between original and reduced TID2013 databases [12] based on objective results

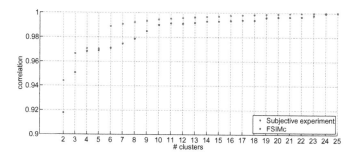

**Fig. 4.** The results of input TID2013 [12] dataset redundancy estimation. '$FSIM_c$' denotes the correlation between full and reduced data, where quality is evaluated by $FSIM_c$ metrics. Subjective experiment means the correlation between full and reduced data, where quality is acquired throughout subjective experiment - forced choice

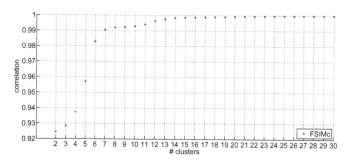

**Fig. 5.** Correlation between original and reduced CISQ databases [3] based on objective results

In the paper [4] the MS-SSIM method was proposed, however the MS-SSIM metrics have much lower correlation values for the most relevant datasets. In the paper the assumption that the metric should characterized by the Pearson's correlations about 0.9 was made. Therefore we used the $FSIM_c$ (Feature Similarity

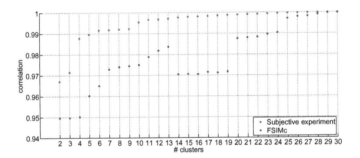

**Fig. 6.** The results of input CISQ [3] dataset redundancy estimation. '$FSIM_c$' denotes the correlation between full and reduced data, where quality is evaluated by $FSIM_c$ metrics. Subjective experiment means the correlation between full and reduced data, where quality is acquired throughout subjective experiment - forced choice

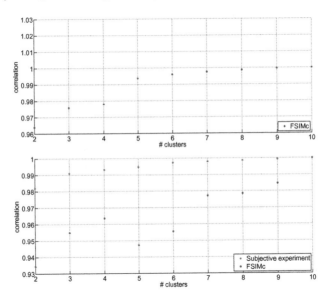

**Fig. 7.** Top: Correlation between original and reduced IVC_SubQualityDB databases [15] based on objective results. Bottom: The results of input IVC_SubQualityDB [15] dataset redundancy estimation. '$FSIM_c$' denotes the correlation between full and reduced data, where quality is evaluated by $FSIM_c$ metrics. Subjective experiment means the correlation between full and reduced data, where quality is acquired throughout subjective experiment - double stimulus

Index Measure) [19] method, as it is reported with high Pearson's correlations between MOS and the objective scores.

We found that the best results, stable with an increasing cluster numbers (starting from 3 clusters), and sample numbers (starting from 12 samples), were received for the forced choice method that, as reported in [4,7], resulted in the smallest measurement variance and thus produced the most accurate results. Promising results were obtained for the double stimulus technique, too. Their correlation level was higher than 90 %.

For practical test of the approach the experiments are performed and analyzed not only on LIVE database [13], but also on the other standard well-known databases: TID2013 [12] (Figs. 3 and 4), CISQ [3] (Figs. 5 and 6) and IVC_SubQualityDB [15] (Fig. 7).

## 5    Conclusions and Future Work

In the paper the efficient approach for subjective experiments was presented. The forced-choice pairwise comparison method was reported to be the most accurate from the tested methods. This method was also found to be the most time-efficient if used in combination with a sorting algorithm that reduces the number of comparisons. The modification of the earlier approach [4], for a scene-preselection is presented. The approach is used to reduce the database size used in subjective experiments and makes it less expensive and therefore more usable. In the place of objective metrics, in the paper we set $FSIM_c$ method instead of MS-SSIM, which have high Pearson correlation with subjective measurements. To verify the procedure, the Pilot stage was computed parallely with two different data: subjective results and objective metrics $FSIM_c$. We check the approach on the well-known databases LIVE [13], IVC_SubQualityDB [15], TID2013 [11] and CISQ [3]. The results obtained with $FSIM_c$ metric are similar as received using MS-SSIM method.

## References

1. Gulliksen, H., Tucker, L.R.: A general procedure for obtaining paired comparisons from multiple rank orders. Psychometrika **26**, 173–184 (1961)
2. Keelan, B.W.: A psychophysical image quality measurement standard. SPIE **5294**, 181–189 (2003)
3. Larson, E.C., Chandler, D.M.: Most apparent distortion: full-reference image quality assessment and the role of strategy. J. Electr. Imaging **19**(1) (2010)
4. Lewandowska (Tomaszewska), A.: Scene reduction for subjective image quality assessment. J. Electr. Imaging **25**(1), 221–226 (2016)
5. (Tomaszewska), A.L.: Time compensation in perceptual experiments. In: Chmielewski, L.J., Kozera, R., Shin, B.-S., Wojciechowski, K. (eds.) ICCVG 2014. LNCS, vol. 8671, pp. 33–40. Springer, Heidelberg (2014). doi:10.1007/978-3-319-11331-9_5
6. Mantiuk, R., Mantiuk, R., Tomaszewska, A., Heidrich, W.: Color correction for tone mapping. Comput. Graph. Forum **28**, 193–202 (2009)
7. Mantiuk, R., Tomaszewska, A., Mantiuk, R.: Comparison of four subjective methods for image quality assessment. Comput. Graph. Forum **31**, 2478–2491 (2012)
8. Pitrey, Y., Barkowsky, M., Pepion, R., Le Callet, P., Hlavacs, H.: Influence of the source content and encoding configuration on the perceived quality for scalable video coding. In: Proceedings of SPIE, vol. 8291, pp. 82911K–82911K-8 (2012)

9. Pinson, M., Wolf, S.: Techniques for evaluating objective video quality models using overlapping subjective data sets [electronic resource]/Pinson, M.H., Wolf, S. U.S. Department of Commerce, National Telecommunications and Information Administration, 1 online resource United States (2008)

10. Pitrey, Y., Robitza, W., Hlavacs, H.: Instance selection techniques for subjective quality of experience evaluation. In: QoEMCS, Part of the EuroITV Conference (2012). http://dcti.iscte.pt/events/qoemcs/

11. Ponomarenko, N., Lukin, V., Zelensky, A., Egiazarian, K., Carli, M., Battisti, F.: TID2008 - a database for evaluation of full-reference visual quality assessment metrics. Adv. Mod. Radioelectronics **10**, 30–45 (2009)

12. Ponomarenko, N., Ieremeiev, L.J.O., Lukin, V., Egiazarian, E., Astola, J., Vozel, B., Chehdi, K., Carli, M., Battisti, F., Jay Kuo, C.C.: Image database TID2013: peculiarities, results and perspectives. Signal Process. Image Commun. **30**, 57–77 (2015)

13. Sheikh, H.R., Sabir, M.F., Bovik, A.C.: A statistical evaluation of recent full reference image quality assessment algorithms. IEEE Trans. Image Process. **15**(11), 3441–3452 (2006)

14. Silverstein, D.A., Farrell, J.E.: Efficient method for paired comparison. J. Electr. Imaging **10**, 394–398 (2001)

15. Strauss, C., Pasteau, F., Autrusseau, F., Babel, M., Bedat, L., Deforges, O.: Subjective and objective quality evaluation of LAR coded art images. In: IEEE International Conference on Multimedia & Expo, ICME 2009, New York, USA (2009)

16. Tomaszewska, A.: Blind noise level detection. In: Campilho, A., Kamel, M. (eds.) ICIAR 2012. LNCS, vol. 7324, pp. 107–114. Springer, Heidelberg (2012). doi:10.1007/978-3-642-31295-3_13

17. Tomaszewska, A.: User study in non-static HDR scenes acquisition. In: Bolc, L., Tadeusiewicz, R., Chmielewski, L.J., Wojciechowski, K. (eds.) ICCVG 2012. LNCS, vol. 7594, pp. 245–252. Springer, Heidelberg (2012). doi:10.1007/978-3-642-33564-8_30

18. Tomaszewska, A., Stefanowski, K.: Real-time spherical harmonics based subsurface scattering. In: Campilho, A., Kamel, M. (eds.) ICIAR 2012. LNCS, vol. 7324, pp. 402–409. Springer, Heidelberg (2012). doi:10.1007/978-3-642-31295-3_47

19. Zhang, L., Zhang, L., Mou, X., Zhang, D.: FSIM: a feature similarity index for image quality assessment. IEEE Trans. Image Process. **20**(8), 2378–2386 (2011)

# Action Recognition Using Silhouette Sequences and Shape Descriptors

Katarzyna Gościewska and Dariusz Frejlichowski[✉]

Faculty of Computer Science and Information Technology, West Pomeranian University of Technology, Żołnierska 52, 71-210 Szczecin, Poland
{kgosciewska,dfrejlichowski}@wi.zut.edu.pl

**Abstract.** The paper provides an approach for human action recognition based on shape analysis. The developed approach is intended for specific type of data, namely sequences of binary silhouettes representing a person performing an action, and consists of several processing steps including shape description as well as similarity or dissimilarity estimation. The approach can deal with sequences of different length without removing any frames. The paper also provides some experimental results showing the classification accuracy and overall recognition effectiveness of the proposed approach using several popular shape description algorithms, namely the Two-Dimensional Fourier Descriptor, Generic Fourier Descriptor, Point Distance Histogram and UNL-Fourier Descriptor.

## 1 Introduction

Human activity recognition has found applications in many areas, especially in video surveillance systems. Nowadays surveillance and security solutions are associated with advanced video content analysis algorithms which support human operators in observation of many scenes and detection of various events related e.g. to abnormal/unusual activities [17,20]. Human activity recognition can be performed on various levels of complexity, depending on the type of activity. According to [18] the simplest level includes recognition of gestures, that is elementary human body movements executed for a short time. In turn, an action is composed of multiple temporarily organized gestures performed by a single person, such as walking, running, bending or waving.

The literature provides a number of shape-based methods that are applied for the recognition of actions using silhouette sequences. For instance, in [11] Trace Transform for each silhouette is extracted, and the whole action sequence is then represented by a final History Trace Template composed of the set of transforms. In [12] an individual silhouette is converted into a one-dimensional representation and then transformed into symbolic vector called SAX (Symbolic Aggregate approXimation). An action is represented by a set of SAX vectors. Some other approaches limit the number of silhouettes and action recognition is based on selected key poses (characteristic frames), e.g. [1,5,15]. Silhouettes can also be accumulated in order to generate spatio-temporal features. A common technique in this category uses motion energy images and motion history images,

© Springer International Publishing AG 2017
R.S. Choraś (ed.), *Image Processing and Communications Challenges 8*,
Advances in Intelligent Systems and Computing 525, DOI 10.1007/978-3-319-47274-4_21

and was proposed in [3]. The basis of the representation is a static vector-image (temporal template) and the vector value at each pixel corresponds to motion properties at that pixel location in the image. The authors of [10] introduced other space-time approach, which utilizes Poisson equation to extract several features, among others, space-time saliency and action dynamics. This approach regards human actions as three-dimensional shapes—accumulated silhouettes in the space-time volume.

This paper focuses solely on action recognition where the amount of processed information about an action is limited to a sequence of binary silhouettes which are then represented using selected shape description algorithms. These representations are subjected to further processing and ultimately are compared on the basis of template matching approach in order to identify overall recognition effectiveness and final classification accuracy for a particular shape descriptor. The rest of the paper is organized as follows: Sect. 2 presents the consecutive steps of the proposed approach and describes the algorithms use for shape representation, Sect. 3 presents some experimental results on action classification and recognition, and Sect. 4 concludes the paper.

## 2 Developed Approach—Data Processing Steps

The developed approach addresses the problem of recognising an action of a single person and uses information contained in a sequence of binary silhouettes. According to [4], the use of silhouettes for action classification assumes that human movement can be represented as a continuous pose change. Then action descriptors can be obtained based on silhouettes extracted from consecutive video frames and traditional classification approaches can be applied. The proposed approach has been already tested using a part of the Weizmann dataset [2]. The original Weizmann dataset contains 90 low-resolution ($180 \times 144$, 50 fps) video sequences of 9 actors performing 10 actions. The corresponding binary masks extracted using background subtraction are available and were used as input data. We have selected five types of actions for the experiments: run (see Fig. 1 for example), walk, bend, jump and one-hand wave. Some data and results will be used in this section to illustrate several processing steps of our approach. Therefore, the following description also includes explanation for the research experiment.

**Fig. 1.** Exemplary silhouettes from a running action sequence (images come from the Weizmann dataset [2])

## 2.1   Step 1—Calculation of Shape Descriptor for Each Silhouette

In this step, each silhouette is represented by one shape descriptor using information about its contour or region. We have already tested four shape description algorithms, namely the Two-Dimensional Fourier Descriptor, Generic Fourier Descriptor, Point Distance Histogram and UNL-Fourier Descriptor. All selected algorithms enable to calculate shape representations of different size and in a form of a vector. It is important due to the fact that we are trying to select the smallest descriptor which simultaneously carries the most information. The selected algorithms have been previously successfully employed for shape analysis and in template matching approaches, e.g. in [7,9], and are described below.

The Two-Dimensional Fourier Descriptor is applied to a region shape. It is calculated as a magnitude of the Fast Fourier Transform and has a form of a matrix with absolute complex values [14]. This algorithm is used as a step in the following two methods. The first one is the UNL-Fourier Descriptor [16]. It is based on contour information and uses centroid to transform Cartesian coordinates of a contour into polar coordinates. New coordinate values are then put into a matrix, where rows represent distances from the centroid and columns the corresponding angles. As a result, an image containing unfolded shape contour in polar coordinates is obtained and then the Two-Dimensional Fourier Descriptor is applied. The second Fourier-based description algorithm is the Generic Fourier Descriptor, which is applied to a region shape and uses the transformation of Cartesian points to polar coordinate system. All pixel coordinates from original region shape image are transformed into polar coordinates and new values are put to a rectangular Cartesian image [19]. Row elements correspond to distances from centroid and the columns to 360 angles. The result has a form of an image and as in the previous case the Two-Dimensional Fourier Descriptor is applied.

The Point Distance Histogram (PDH) is a shape descriptor that utilizes information about shape contour [8]. In order to derive a PDH representation, an origin of the polar transform of a contour is firstly selected, usually a centroid. Polar coordinates are stored in two vectors—one for angles and one for radii. In the next step, angle values are converted to the nearest integers. Then the elements in both vectors are rearranged with respect to the increasing angle values. If any equal angles exist, then only the element with the highest radii value is left. Next, only radii vector is selected for further processing. Its elements are normalized and assigned to bins in the histogram. Then values in histogram bins are normalized according to the highest one and the final representation is obtained.

## 2.2   Step 2—Matching All Shape Descriptors Within One Sequence

For a given sequence, this step includes calculation of dissimilarities between first frame and the rest of frames using Euclidean distance. The resulting vector containing distance values, normalized to interval [0, 1], is a one-dimensional descriptor of a sequence—a distance vector. The number of its elements equals the number of silhouettes in the input sequence. Figure 2 shows exemplary plots

**Fig. 2.** Exemplary plots of distance vectors for three actions performed by the same actor: bend, walk and run respectively. Low peaks correspond to the silhouettes that are most similar to the first silhouette in a sequence. The exemplary distance vectors were obtained using $2 \times 2$ subpart of the Two-Dimensional Fourier Descriptor. For walk and run actions the periodicities are noticeable

of distance vectors of three different actions. The plots reveal action periodicities and the differences between actions.

### 2.3   Step 3—Converting the Distance Vectors into Sequence Representations

The next step aims to convert distance vectors into the form and size that enables the calculation of similarity between them. Therefore, distance vectors were treated as signals and it turned out that the best way to transform such a signal was to use a periodogram. Periodogram is a spectral density estimation of a signal and it can determine hidden periodicities in data [6]. In most cases, the results have improved when fast Fourier transform was firstly applied to a distance vector, the magnitude was extracted and after that the periodogram was used. Due to the fact that distance vectors varied in size, the periodogram helped to equalize final representations' sizes. Ultimately, one periodogram represents one silhouette sequence.

### 2.4   Step 4—Selection of Matching Procedure and Splitting Data

This step solves the problem of how to split data into templates and test objects. Some initial tests showed that final results were dependent on which part of the data was selected as templates (one template is a single class representative). Therefore, being inspired by the k-fold cross-validation technique [13] we have decided to perform the experiment several times using different set of templates in each iteration. The final recognition effectiveness is then the average of the results from all iterations. For instance, the first iteration used objects with numbers from 1 to $k$ as templates and objects with numbers from $k+1$ to $n$ as test objects, then the second iteration used objects with numbers from $k+1$ to $2*k$ as templates and the rest as test objects, and so on.

## 2.5   Step 5—Estimation of the Similarity Between Sequence Descriptors

This step includes the calculation of similarity between sequence descriptors using template matching approach and variable template set as described in the previous step. The correlation coefficient is used. Here template matching, for one iteration, is understood as a process that compares each test object with all templates and selects the most similar one, what simultaneously indicates the probable class of a particular test object. Then the results can be interpreted and analysed in three different ways (considering only the number of correct classifications—'true positive'):

1. Overall recognition effectiveness for each shape descriptor, averaged for all classes and all iterations,
2. Classification accuracy for each iteration, each shape descriptor and all classes,
3. Classification accuracy for each class, each shape descriptor and all iterations.

# 3   Experiments and Results

Several experiments have been carried out in order to verify the effectiveness and accuracy of the proposed approach. Each experiment consisted of five steps described in Sect. 3, except that for each experiment different shape description algorithm was used. Moreover, various size of shape representation were employed, namely the $2 \times 2$, $5 \times 5$, $10 \times 10$, $25 \times 25$ and $50 \times 50$ absolute spectrum subparts for the Two-Dimensional Fourier Descriptor, Generic Fourier Descriptor and UNL-Fourier Descriptor, and 2, 5, 10, 25, 50, 75 and 100 histogram bins for the Point Distance Histogram. In the experimental database there were 45 silhouette sequences of 9 actors performing 5 actions—bend, jump, run, walk and one-hand wave—taken from the Weizmann dataset [2]. The number of frames (silhouettes) in a sequence varied from 28 to 125. During the experiment each subgroup of 5 sequences of one person performing these actions was iteratively used as a template set. Percentage experimental results are presented below with respect to the three result analysis manners (see Step 5. of the approach).

Average recognition effectiveness values for each shape descriptor, all classes and all iterations are as follows:

- 49.2 %, 46.1 %, 42.8 %, 41.1 % and 39.4 % for the $2 \times 2$, $5 \times 5$, $10 \times 10$, $25 \times 25$ and $50 \times 50$ subparts of the Two-Dimensional Fourier Descriptor respectively;
- 51.7 %, 51.4 %, 51.7 %, 52.2 % and 50.0 % for the $2 \times 2$, $5 \times 5$, $10 \times 10$, $25 \times 25$ and $50 \times 50$ subparts of the Generic Fourier Descriptor respectively;
- 36.7 %, 39.7 %, 37.8 %, 36.9 % and 39.7 % for the $2 \times 2$, $5 \times 5$, $10 \times 10$, $25 \times 25$ and $50 \times 50$ subparts of the UNL-Fourier Descriptor respectively;
- 39.7 %, 36.7 %, 36.1 %, 34.7 %, 34.4 %, 34.4 % and 34.4 % for the 2, 5, 10, 25, 50, 75 and 100 histogram bins of the Point Distance Histogram respectively.

**Table 1.** Results for the experiment using Generic Fourier Descriptor—classification accuracy for each iteration, each shape descriptor and averaged for all classes

| Iteration | Descriptor size | | | | |
|---|---|---|---|---|---|
| No | $2 \times 2$ | $5 \times 5$ | $10 \times 10$ | $25 \times 25$ | $50 \times 50$ |
| 1 | 45.0% | 45.0% | 45.0% | 50.0% | 47.5% |
| 2 | 37.5% | 37.5% | 35.0% | 35.0% | 35.0% |
| 3 | 52.5% | 52.5% | 52.5% | 52.5% | 52.5% |
| 4 | 52.5% | 52.5% | 50.0% | 52.5% | 37.5% |
| 5 | 45.0% | 42.5% | 40.0% | 45.0% | 47.5% |
| 6 | 47.5% | 47.5% | 50.0% | 50.0% | 50.0% |
| 7 | 57.5% | 57.5% | 60.0% | 57.5% | 55.0% |
| **8** | **67.5%** | **67.5%** | **72.5%** | **70.0%** | **60.0%** |
| 9 | 60.0% | 60.0% | 60.0% | 57.5% | 65.0% |

The average results indicate the Generic Fourier Descriptor as the most effective shape description algorithm for the employed approach and selected data. The percentage recognition effectiveness values are similar for all descriptor sizes. Additional results will enable for a more detailed insight into the classification accuracy using the Generic Fourier Descriptor (see Tables 1, 2 and 3). Table 1 illustrates the results obtained in each iteration and averaged for all classes. It can be seen that the classification accuracy values vary between iterations and that the best result is obtained in iteration no. 8. This can be interpreted in such a way that templates used in this iteration are represented by the most distinctive features enabling proper class indication.

In Table 2, the averaged results for all iterations can be found. It can be clearly seen that 'bend' action is the most distinctive one, while the 'jump' action is the least recognizable. It is not obvious which shape description size should be indicated as the best to employ for shape representation, because it varies depending on the class. However, if only iteration no. 8 is taken under

**Table 2.** Results for the experiment using Generic Fourier Descriptor—classification accuracy for each class, each shape descriptor and averaged for all iterations

| Class | Descriptor size | | | | |
|---|---|---|---|---|---|
| | $2 \times 2$ | $5 \times 5$ | $10 \times 10$ | $25 \times 25$ | $50 \times 50$ |
| 'bend' | 87.5% | 87.5% | 88.9% | 88.9% | 87.5% |
| 'jump' | 27.8% | 26.4% | 30.6% | 31.9% | 27.8% |
| 'run' | 48.6% | 50.0% | 51.4% | 54.2% | 55.6% |
| 'walk' | 38.9% | 37.5% | 43.1% | 43.1% | 44.4% |
| 'wave' | 55.6% | 55.6% | 44.4% | 43.1% | 34.7% |

**Table 3.** Classification accuracy for each class, various size of Generic Fourier Descriptor and iteration no. 8

| Class | Descriptor size | | | | |
|---|---|---|---|---|---|
| | $2 \times 2$ | $5 \times 5$ | $\mathbf{10 \times 10}$ | $25 \times 25$ | $50 \times 50$ |
| 'bend' | 100 % | 100 % | **100 %** | 100 % | 100 % |
| 'jump' | 25.0 % | 25.0 % | **37.5 %** | 37.5 % | 25.0 % |
| 'run' | 87.5 % | 87.5 % | **87.5 %** | 87.5 % | 75.0 % |
| 'walk' | 62.5 % | 62.5 % | **75.0 %** | 75.0 % | 75.0 % |
| 'wave' | 62.5 % | 62.5 % | **62.5 %** | 50.0 % | 25.0 % |

consideration—due to the most distinctive templates—it turns out that the proposed approach is most effective when the $10 \times 10$ subpart of the Generic Fourier Descriptor is used. Table 3 depicts percentage classification accuracy values for iteration no. 8.

## 4  Summary and Conclusions

In the paper, an approach for action recognition based on silhouette sequences has been presented. It uses various shape description algorithms to represent silhouettes and Euclidean distance to estimate dissimilarity between shape descriptors within a sequence. Normalized distances create a vector representation of the sequence. All sequence representations are further processed using fast Fourier transform and periodogram, and ultimately are compared using template matching approach and correlation coefficient. Experimental results showed that the developed approach is most effective and accurate when the Generic Fourier Descriptor is used. Generally, the results are promising, however the developed approach needs further improvements and should be examined using more data. Future works involve experimental verification of other shape descriptors and matching measures, that will make the approach more effective.

## References

1. Baysal, S., Kurt, M.C., Duygulu, P.: Recognizing human actions using key poses. In: 20th International Conference on Pattern Recognition, pp. 1727–1730, August 2010
2. Blank, M., Gorelick, L., Shechtman, E., Irani, M., Basri, R.: Actions as space-time shapes. In: The Tenth IEEE International Conference on Computer Vision, pp. 1395–1402 (2005)
3. Bobick, A.F., Davis, J.W.: The recognition of human movement using temporal templates. IEEE Trans. Pattern Anal. Mach. Intell. **23**(3), 257–267 (2001)
4. Borges, P.V.K., Conci, N., Cavallaro, A.: Video-based human behavior understanding: a survey. IEEE Trans. Circ. Syst. Video Technol. **23**(11), 1993–2008 (2013)

5. Chaaraoui, A.A., Climent-Pérez, P., Flórez-Revuelta, F.: Silhouette-based human action recognition using sequences of key poses. Pattern Recogn. Lett. **34**(15), 1799–1807 (2013)
6. Chitode, J.: Digital Signal Processing. Technical Publications, Pune (2009)
7. Forczmański, P., Frejlichowski, D.: Robust stamps detection and classification by means of general shape analysis. In: Bolc, L., Tadeusiewicz, R., Chmielewski, L.J., Wojciechowski, K. (eds.) ICCVG 2010. LNCS, vol. 6374, pp. 360–367. Springer, Heidelberg (2010). doi:10.1007/978-3-642-15910-7_41
8. Frejlichowski, D.: An experimental comparison of three polar shape descriptors in the general shape analysis problem. In: Swiatek, J., Borzemski, L., Grzech, A., Wilimowska, Z. (eds.) Information Systems Architecture and Technology – System Analysis in Decision Aided Problems, pp. 139–150. Oficyna Wydawnicza Politechniki Wrocławskiej (2010)
9. Frejlichowski, D.: Pre-processing, extraction and recognition of binary erythrocyte shapes for computer-assisted diagnosis based on mgg images. In: Bolc, L., Tadeusiewicz, R., Chmielewski, L., Wojciechowski, K. (eds.) Computer Vision and Graphics, pp. 368–375. Springer, Berlin (2010)
10. Gorelick, L., Blank, M., Shechtman, E., Irani, M., Basri, R.: Actions as space-time shapes. Trans. Pattern Anal. Mach. Intell. **29**(12), 2247–2253 (2007)
11. Goudelis, G., Karpouzis, K., Kollias, S.: Exploring trace transform for robust human action recognition. Pattern Recogn. **46**(12), 3238–3248 (2013)
12. Junejo, I.N., Junejo, K.N., Aghbari, Z.A.: Silhouette-based human action recognition using sax-shapes. Vis. Comput. **30**(3), 259–269 (2014)
13. Kohavi, R.: A study of cross-validation and bootstrap for accuracy estimation and model selection. In: Proceedings of the 14th International Joint Conference on Artificial Intelligence, IJCAI 1995, vol. 2, pp. 1137–1143. Morgan Kaufmann Publishers Inc., San Francisco (1995)
14. Kukharev, G.: Digital Image Processing and Analysis (in Polish). SUT Press, Szczecin (1998)
15. Liu, L., Shao, L., Zhen, X., Li, X.: Learning discriminative key poses for action recognition. IEEE Trans. Cybern. **43**(6), 1860–1870 (2013)
16. Rauber, T.W.: Two dimensional shape description. Technical report, Universidade Nova de Lisboa, Lisoba, Portugal (1994)
17. Vaswani, N., Roy-Chowdhury, A.K., Chellappa, R.: Shape activity: a continuous-state hmm for moving/deforming shapes with application to abnormal activity detection. IEEE Trans. Image Process. **14**(10), 1603–1616 (2005)
18. Vishwakarma, S., Agrawal, A.: A survey on activity recognition and behavior understanding in video surveillance. Vis. Comput. **29**(10), 983–1009 (2012)
19. Zhang, D., Lu, G.: Shape-based image retrieval using generic fourier descriptor. Signal Process. Image Commun. **17**(10), 825–848 (2002)
20. Zhong, H., Shi, J., Visontai, M.: Detecting unusual activity in video. In: Proceedings of the 2004 IEEE Computer Society Conference on Computer Vision and Pattern Recognition, vol. 2, pp. II-819-II-826, June 2004

# Influence of Aggregating Window Size on Disparity Maps Obtained from Equal Baseline Multiple Camera Set (EBMCS)

Adam L. Kaczmarek[✉]

Faculty of Electronics, Telecommunications and Informatics,
Gdansk University of Technology, ul. G. Narutowicza 11/12, 80-233 Gdansk, Poland
adam.l.kaczmarek@eti.pg.gda.pl

**Abstract.** This paper is concerned with obtaining disparity maps on the basis of images from Equal Baseline Multiple Camera Set (EBMCS). EBMCS consists of a central camera and side cameras. Algorithms for obtaining disparity maps with the use of EBMCS take advantage of aggregating windows similarly to stereo matching algorithms for a stereo camera, a camera matrix or a camera array. The paper analyzes the influence of aggregating window size on the quality of disparity maps. Experiments presented in this paper include Sum of Sum of Squared Differences (SSSD) and Sum of Sum of Absolute Differences (SSAD) matching cost functions. Results show that for EBMCS with five cameras the highest quality of disparity maps is obtained when the size of the aggregating window is on average over 55 % smaller than the size of the most effective window for a pair of cameras.

## 1 Introduction

There is a variety of devices designed for estimating distances to objects. A list of such equipment includes time-of-flight cameras (TOF) [6], Light Detection and Ranging (LIDAR) [13], structured-light 3D scanners (such as Microsoft Kinect) [2] and cameras [4]. All these tools, apart from cameras, perform measurements by emitting a light beam and then recording its reflection. This kind of a distance estimation method has a major disadvantage such that the intensity of emitted light needs to be high enough to illuminate an examined object. The necessary intensity depends on the existing light conditions and the distance from a device to an object [1,6]. This problem does not apply to cameras which are useful in both highly illuminated environments and when objects are located at a large distance from a measuring device.

Typical equipment for this method of measuring is a stereo camera which consists of two cameras aimed in the same direction. The same object visible from both cameras is placed at different locations in a pair of images from these cameras. Images from stereo cameras are processed by stereo matching algorithms in order to reveal these disparities. A set of disparities for objects and their parts visible in images forms a disparity map. Considering the distance

R.S. Choraś (ed.), *Image Processing and Communications Challenges 8*,
Advances in Intelligent Systems and Computing 525, DOI 10.1007/978-3-319-47274-4_22

between cameras and focal lengths of their lens disparity maps can be converted to depth maps containing values of distances [9].

Disparity maps can be also obtained with the use of more than two cameras. Camera arrays, camera matrices and other arrangements of cameras are used [9,14]. This paper is concerned with obtaining disparity maps with the use of a set of cameras called by the author of this paper Equal Baseline Multiple Camera Set (EBMCS). The set consists of up to five cameras. In the set there is a central camera and side cameras located around the central one [4].

This paper examines the issue of selecting for EBMCS the size of an aggregating window which makes it possible to obtain the highest quality of disparity maps. Aggregating windows are used in algorithms for obtaining disparity maps by comparing parts of images from different cameras in order to detect areas corresponding to the same viewed object. Different sizes of windows are suitable for stereo cameras and multiple camera sets.

## 2    Related work

Okutomi and Kanade are authors of one of the most significant papers in the field of multi-camera vision systems [9]. They described a method for obtaining depth maps on the basis of images from a camera array. The array was considered as a set of stereo cameras consisting of the first camera and some other camera in the array. Each stereo camera has a different baseline that is a distance between cameras. As a result, in different stereo cameras values of disparities are different for the same object visible from these cameras. Okutomi and Kanade solved this problem by using inverse distances from the array to viewed objects instead of disparities.

Camera arrays are also used for purposes other than estimating distances. A set of images from an array makes it possible to remove from images an object visible in the foreground of a viewed scene. The removal results in replacing the view of this object with the view of other object located behind the removed one. Pei et al. performed such modifications on images containing views of people [11]. The author of this paper also analyzed this kind of operations with images of buildings and their surroundings [5].

Another kind of a camera arrangement was used by Park and Inoue [10]. They determined a reference camera and four side cameras equally distant from the central one. The same kind of a camera arrangement was used in the research presented in this paper. Methods for obtaining disparity maps from this set were also proposed by the author of this paper [4].

Hensler et al. used a central cameras and side cameras similarly as Park and Inoue [3]. However, their set contained three side cameras. Locations of cameras corresponded to vertices of an equilateral triangle. Moreover, Nalpantidis described a quad-camera system arranged in a form of a matrix [8]. Sets with greater number of cameras were also used. Wilburn et al. experimented with a set containing 100 cameras [14]. Sensors were arranged in a form of a camera matrix which consisted of parallel camera arrays. Wilburn et al. used the

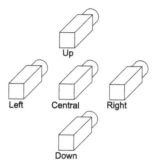

**Fig. 1.** The arrangement of cameras in Equal Baseline Multiple Camera Set

matrix for making images of fast moving objects. Matusik used a camera matrix consisting of 16 cameras for making 3D videos [7].

## 3   EBMCS Camera Arrangement

In the experiments presented in this paper, Equal Baseline Multiple Camera Set is used. The set consists of a central camera and up to four cameras located around the central one. The set containing five cameras is the same kind of a camera set as the one used by Park and Inoue [10]. There is a left, a right, an up and a down side camera. All cameras are placed in the same plane and aimed in the same direction. Distances between each side camera and a central camera are the same.

The central camera provides an image which is a reference one. Points of this image correspond to points of a disparity map obtained with the use of the described set. Images from other cameras are used for determining values of points in the resulting disparity maps. The arrangement of cameras in EBMCS is presented in Fig. 1. In order to evaluate EBMCS, experiments were performed with two data sets. Data sets were prepared by making images of plants with the use of EBMCS containing five cameras [4]. Sets consisted of images of Dwarf Umbrella Tree (Fig. 2) and images of Ficus Tree (Fig. 3).

## 4   Matching Measures for EBMCS

The previous research performed by the author of this paper analyzed the performance of different algorithms for obtaining disparity maps with the use of EBMCS [4]. The research resulted in developing the Similar Areas Matching algorithm dedicated to specifics of EBMCS. This algorithm does not use aggregating windows which are the subject of this paper. The previous research was also concerned with analyzing results of adapting different matching measures for the purpose of obtaining disparity maps with the use of EBMCS.

All matching measures estimate levels of differences between some areas of images. The size of areas is defined in a form of an aggregating window. A stereo

(a) up        (b) left        (c) central        (d) right        (e) down

**Fig. 2.** Images of Dwarf Umbrella Tree [4]

(a) up        (b) left        (c) central        (d) right        (e) down

**Fig. 3.** Images of Ficus Tree [4]

matching algorithm compares points within the aggregating window in the reference image with points in the aggregating window located at corresponding coordinates in the side image. Obtaining disparities with the use of EBMCS is performed similarly as in case of processing images from a stereo camera. However, instead of comparing aggregating windows in two images these windows are simultaneously matched in all images included in the set.

Different measures use different methods of estimating levels of differences between parts of images covered by aggregating windows. The research performed with EBMCS showed that using SSSD (Sum of Sum of Squared Differences) (Eq. 1) and SSAD (Sum of Sum of Absolute Differences) (Eq. 2) measures generated the best results [4].

$$SSAD\left(\mathbf{p}, \mathbf{d}\right) = \sum_{1 \leq i \leq N} \sum_{\mathbf{b} \in \mathbf{W}} |I_0(\mathbf{p} + \mathbf{b}) - I_i(\mathbf{p} + \mathbf{b} + \mathbf{d_i})| \tag{1}$$

$$SSSD\left(\mathbf{p}, \mathbf{d}\right) = \sum_{1 \leq i \leq N} \sum_{\mathbf{b} \in \mathbf{W}} (I_0(\mathbf{p} + \mathbf{b}) - I_i(\mathbf{p} + \mathbf{b} + \mathbf{d_i}))^2 \tag{2}$$

where SSAD and SSSD are values of measures, $\mathbf{p}$ is the point for which disparity is calculated, $\mathbf{d}$ is a disparity, $N$ is total number of points, $\mathbf{W}$ is an aggregating window, $I_i$ is the intensity of a point in an image from camera $i$ and $\mathbf{d_i}$ is a disparity in camera $i$. The central camera has index 0, side cameras have indexes in the range from 1 to 4.

## 5    Quality Measure and Ground Truth

Disparity maps obtained from EBMCS were evaluated with the use of the percentage of bad matching pixels metric (BMP). The metric was described by

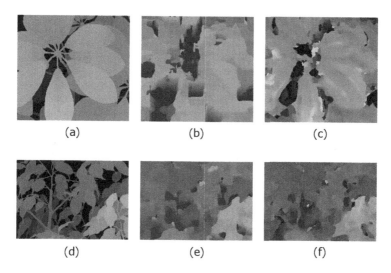

**Fig. 4.** Dwarf Umbrella Tree: (a) Ground truth (b) Disparity map, 2 cameras, aggregating window 14 (c) Disparity map, 5 cameras, aggregating window 6; Ficus Tree: (d) Ground truth (e) Disparity map, 2 cameras, aggregating window 9 (f) Disparity map, 5 cameras, aggregating window 5.

Scharstein and Szeliski [12]. The formula for BMP is presented in Eq. 3.

$$BMP = \frac{1}{N} \sum_{\mathbf{p}} (|D_M(\mathbf{p}) - D_T(\mathbf{p})| > Z) \qquad (3)$$

where $BMP$ is value of the metric, $Z$ is the disparity error tolerance, $N$ is the number of considered points, $D_M(\mathbf{p})$ is the disparity of the point $\mathbf{p}$ located in the disparity map and $D_T(\mathbf{p})$ is the disparity in ground truth at the same coordinates. In the experiments presented in this paper, the error tolerance level was set to 4. Moreover, the BMP measure was calculated for all point for which ground truth contained disparities.

Ground truth was prepared for both data sets used in the experiments. Ground truth data is presented in Fig. 4a (Dwarf Umbrella Tree) and Fig. 4d (Ficus Tree). Intensities of points in Fig. 4 correspond to values of disparities. The brighter is a point, the greater is its disparity.

Ground truth does not contain disparities for all points of the image. In some areas of the image it was not possible to determine real values of disparities because these points correspond to object located at the background of a viewed scene or in border parts of the image [4].

## 6   Experiments

Experiments were performed with two data sets presented in Sect. 3. Results for both SSAD and SSSD matching measures are presented in Fig. 5. Charts presented in figures refer to four different configurations of EBMCS. Configurations

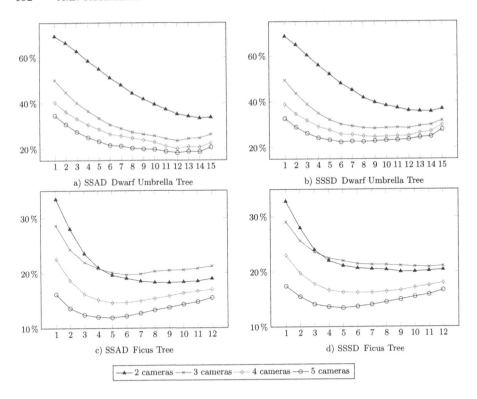

**Fig. 5.** Results of SSAD and SSSD measures with respect to aggregating window size

differ in the number of cameras included in EBMCS in the range between two to five.

Vertical axes of charts show values of percentage of bad matching pixels metric (Sect. 5). Horizontal axes correspond to sizes of aggregating windows. In the experiments, square aggregating windows were used. The size $r$ is the number of points between the centre of the square and its border. The area of an aggregating window is equal to $(2r + 1)^2$ points.

In case of using a pair of cameras for images of Dwarf Umbrella Tree the highest quality of disparity maps is obtained when the size of aggregating window is equal to 14. The disparity map generated with this configuration and the SSSD measure is presented in Fig. 4b. With the use of greater number of cameras, the size of a matching window can be significantly reduced without decreasing the quality of maps. EBMCS with five cameras and aggregating window equal to only 1 generates disparity maps with the better quality than any window size used with a pair of cameras. Moreover, with the use of a number of cameras greater than 2, the quality of disparity map can be further improved by using larger aggregating windows. In case of using the SSAD measure for the Dwarf Umbrella Tree set, the best maps are acquired when windows size is 12. With the use of the SSSD measure the most advantageous size of windows varies between

6 and 9. Differences in values of BMP are not significant (lower than 1.86 %) for the same number of cameras used with window sizes in this range. The best results for SSSD are obtained with the use of five cameras and the aggregating window equal to 6. The disparity map obtained for these parameters is presented in Fig. 4c.

In case of using two cameras for images of Ficus Tree the best results are obtained when the window size is equal to 9. However, increasing the number of cameras to three does not cause the increase in the quality of disparity maps when windows greater than 4 are used. This configuration of cameras is ineffective for this set of images. Nevertheless, results improve when four and five cameras are used. Using five cameras with any kind of considered window sizes leads to better results than using two cameras. The best results for five cameras are obtained when the window size is 5 for both SSAD and SSSD. Figure 4e presents results of using two cameras with SSSD and the aggregating window equal to 9 for images of Ficus Tree. The result for five cameras, the SSSD measure and the window equal to 5 is shown in Fig. 4f.

## 7  Summary

The main advantage of using Equal Baseline Multiple Camera Set is the possibility to acquire disparity maps which have a higher quality than maps obtained with the use of a single stereo camera. Moreover, the camera set can be applied in conditions in which other kinds of devices for estimating distances cannot be used. TOF cameras and structured light 3D scanners (including Microsoft Kinect camera) does not provide accurate data when distances are measured to objects which are illuminated by intensive natural light in an out-door environment. These devices also cannot precisely measure distances to large objects such as buildings.

An algorithm for obtaining disparity maps from EBMCS needs to process more images than a matching algorithm for a stereo camera. It may seem that this leads to the increase in the number of required calculations. However, with the use of EBMCS the size of aggregating windows can be reduced. In case of obtaining disparity maps with the highest quality the most advantageous window size for five cameras is, on average, equal to 7 for the analyzed data sets, while average size of the best window for stereo cameras is 11.5. The area of the rectangle window in the first case is equal to 255 (Sect. 6), the area in the second case is on average 576. Therefore, the size of the best window for five cameras is more than 55 % lower than the best window for a stereo camera. Moreover, even for the smallest window size EBMCS produces better results than stereo camera with a large aggregating window. With the use of EBMCS the quality of disparity maps can be improved without the necessity to increase the number of calculations. Experiments were performed for algorithms based on SSAD and SSSD measures, however in further work the author is planning to verify the usability of EBMCS with different kinds of algorithms for obtaining disparity maps.

# References

1. Gupta, M., Yin, Q., Nayar, S.K.: Structured light in sunlight. In: 2013 IEEE International Conference on Computer Vision, pp. 545–552 (2013)
2. Ha, H., Oh, T.H., Kweon, I.S.: A multi-view structured-light system for highly accurate 3D modeling. In: International Conference on 3D Vision (3DV), pp. 118–126 (2015)
3. Hensler, J., Denker, K., Franz, M., Umlauf, G.: Hybrid face recognition based on real-time multi-camera stereo-matching. In: Bebis, G., Boyle, R., Parvin, B., Koracin, D., Wang, S., Kyungnam, K., Benes, B., Moreland, K., Borst, C., DiVerdi, S., Yi-Jen, C., Ming, J. (eds.) ISVC 2011. LNCS, vol. 6939, pp. 158–167. Springer, Heidelberg (2011). doi:10.1007/978-3-642-24031-7_16
4. Kaczmarek, A.L.: Improving depth maps of plants by using a set of five cameras. J. Electron. Imaging 24(2), 023018:1–023018:11 (2015)
5. Kaczmarek, A.L.: Widzenie komputerowe oparte na mnogości widoków. Zeszyty Naukowe Wydziału Elektrotechniki i Automatyki Politechniki Gdańskiej 36, 85–88 (2013)
6. Kazmi, W., Foix, S., Alenya, G.: Plant leaf imaging using time of flight camera under sunlight, shadow and room conditions. In: IEEE International Symposium on Robotic and Sensors Environments (ROSE), pp. 192–197 (2012)
7. Matusik, W., Pfister, H.: 3D TV: a scalable system for real-time acquisition, transmission, and autostereoscopic display of dynamic scenes. In: ACM SIGGRAPH, pp. 814–824 (2004)
8. Nalpantidis, L., Chrysostomou, D., Gasteratos, A.: Obtaining reliable depth maps for robotic applications from a quad-camera system. In: Xie, M., Xiong, Y., Xiong, C., Liu, H., Hu, Z. (eds.) ICIRA 2009. LNCS (LNAI), vol. 5928, pp. 906–916. Springer, Heidelberg (2009). doi:10.1007/978-3-642-10817-4_89
9. Okutomi, M., Kanade, T.: A multiple-baseline stereo. IEEE Trans. Pattern Anal. Mach. Intell. 15(4), 353–363 (1993)
10. Park, J.I., Inoue, S.: Acquisition of sharp depth map from multiple cameras. Sig. Process. Image Commun. 14(1–2), 7–19 (1998)
11. Pei, Z., Zhang, Y., Yang, T., Zhang, X., Yang, Y.H.: A novel multi-object detection method in complex scene using synthetic aperture imaging. Pattern Recogn. 45(4), 1637–1658 (2012)
12. Scharstein, D., Szeliski, R.: A taxonomy and evaluation of dense two-frame stereo correspondence algorithms. Int. J. Comput. Vision 47(1), 7–42 (2002)
13. Tse, R.O., Gold, C., Kidner, D.: 3D city modelling from LIDAR data. In: van Oosterom, V., Zlatanova, S., Penninga, F., Fendel, E.M. (eds.) Advances in 3D Geoinformation Systems. Lecture Notes in Geoinformation and Cartography, pp. 161–175. Springer, Berlin (2008)
14. Wilburn, B., Joshi, N., Vaish, V., Talvala, E.V., Antunez, E., Barth, A., Adams, A., Horowitz, M., Levoy, M.: High performance imaging using large camera arrays. In: ACM SIGGRAPH, pp. 765–776 (2005)

# Combining Image Thresholding and Fast Marching for Nuclei Extraction in Microscopic Images

Marek Kowal[(✉)], Przemysław Jacewicz, and Józef Korbicz

Institute of Control and Computation Engineering,
Univeristy of Zielona Góra, Zielona Góra, Poland
{M.Kowal,P.Jacewicz,J.Korbicz}@issi.uz.zgora.pl

**Abstract.** Computer-Aided Diagnosis (CAD) in digital pathology very often boils down to examination of nuclei using morphological analysis. To determine the characteristics of nuclei, they need to be segmented from the background or other objects in the image (e.g. red blood cells). Despite a tremendous work that has been done to improve segmentation methods, nuclei segmentation remains a very challenging problem. This particularly applies to the cytological images, where nuclei often touch, overlap, cluster, are obscured, or destroyed. Most well known methods of image processing cannot cope with this challenge. Nevertheless, in this study we demonstrated that methods like image thresholding, edge detection, erosion and fast marching, when combined, give satisfactory segmentation results. The proposed approach uses isodata image thresholding and Canny edge detection to find nuclei regions in the image. Then this information is employed to determine centers of the nuclei using conditional erosion. Finally, fast marching algorithm extracts nuclei. The method was applied to extract nuclei from microscopic images of cytological material obtained from breast. Different morphometric, textural, colorimetric and topological features were computed for segmented nuclei to describe cases (patients). The effectiveness of the segmentation was evaluated in terms of classification accuracy of breast cancer, where the cases were classified as either benign or malignant. The acquired predictive accuracy was 98 %, which is very promising and shows that the presented method ensures accurate nuclei segmentaion in cytological images.

**Keywords:** Nuclei segmentation · Image thresholding · Fast marching · Breast cancer · Cytology

## 1 Introduction

Studies and diagnoses diseases by analysis of microscopic images of cells have been a goal of human pathology since the middle of the 19th century. Recent advances in this field have revolutionized pathology by the introduction of digital microscopy. Nowadays, more and more pathologists examine the biological material on a computer screen instead of reviewing it under a microscope. But still,

R.S. Choraś (ed.), *Image Processing and Communications Challenges 8*,
Advances in Intelligent Systems and Computing 525, DOI 10.1007/978-3-319-47274-4_23

they must devote many years to gain experience to become specialist in the identification of cancer cells. Along with the development of advanced vision systems and computer science, quantitative cytopathology can support the pathologist's work by the automation of nuclei segmentation, cell population count, computing statistics of morphological features or cell classification. But to accomplish these tasks we need accurate algorithm for cell or nucleus segmentation. If we look at cytological specimen under a microscope, we will see a lot of clumps of nuclei which create complex, random and heterogeneous structures without clear boundaries. Unfortunately, their segmentation is rather a challenging task.

Many publications present the possibility of using watershed methods, graph cuts, level sets, image thresholding and data clustering based methods for nucleus segmentation [3–7]. However, the problem is still open because there are no universal segmentation methods that can be applied for images representing biological material from different tissues.

In this work we limit our researches and discussion to the issue of nuclei segmentation in cytological images of biological material obtained from breast using fine needle biopsy (FNB). We propose a two-stage procedure to extract nuclei from the image. Firstly, isodata based image thresholding is used to background-foreground segmentation. Then, Canny edge detector is applied to highlight boundaries between the nuclei in foreground region. Nonetheless, separation of nuclei is still not perfect. Thus, conditional erosion is used to find the centers of the nuclei. They are used to initialize the fast marching algorithm which extracts individual nuclei. The effectiveness of nuclei segmentation is verified in terms of classification accuracy of breast cancer malignancy. This is done by using morphometric, colorimetric, textural and topological features of nuclei.

The remainder of this paper is organized as follows. In Sect. 2, Materials and Methods are described. Section 3 gives the description of the experiments and study of the results obtained for proposed method. Concluding remarks are given in Sect. 4.

## 2    Materials and Methods

### 2.1    Study Dataset

The first step of the diagnostic process begins with the acquisition of biological material from the breast. The cytological material was obtained by FNB from 50 patients of the Regional Hospital in Zielona Góra, Poland. The set contains 25 benign and 25 malignant lesions cases. Smears from the biological material were fixed in spray fixative and dyed with hematoxylin and eosin. Cytological preparations were then digitalized into virtual slides using the Olympus VS120 Virtual Microscopy System. Virtual slides offer a radically improved level of detail in comparison to the images grabbed with analog camera. One can clearly observe clumps of chromatin (especially in malignant cases), nucleoli, and nuclear membranes. Next, on each slide, a pathologist selected 11 distinct areas which were converted to 8 bit/channel RGB TIFF files the size of 1583 × 828 pixels. The number of areas per one patient was recommended by the pathologists at the hospital and allows for a correct diagnosis. The image database contains 550 images (11 images per patient). Both malignant and benign sets contain the same number of images (275 images describing 25 cases). All cancers were histologically confirmed and all patients with benign disease were either biopsied or followed for a year.

## 2.2   Nuclei Segmentation

The whole procedure starts from converting original image $I$ to the binary image $BW$ with nuclei region highlighted. Binary image $BW$ is the result of image thresholding and edge detection [9,11]. Threshold $\theta$ is calculated iteratively using isodata procedure [8]. The histogram is initially divided into two parts using a starting threshold $\theta_0$ such as half of the maximum dynamic range. The sample mean $m_{f0}$ of the gray values associated with the foreground (nuclei) pixels and the sample mean $m_{b,0}$ of the gray values associated with the background pixels are determined. Then, new threshold value $\theta_k$ is computed as the average of these two sample means. The process of computing new threshold continues, until the threshold does not change any more. This procedure is repeated for each image separately. Unfortunately, image thresholding procedure is not able to extract individual nuclei from densely packed clumps Fig. 1(b). To overcome this problem we propose to use Canny edge detection algorithm to find the nuclei borders. Edges are mapped on a results of thresholding in order to cut individual nuclei from clumps. It can be observed that on this stage of image segmentation some nuclei are properly segmented but there are also some nuclei that are stuck together Fig. 1(c). Further processing is necessary to obtain more nuclei for analysis.

A key stage of proposed segmentation procedure is to correctly mark nuclei centers to seed fast marching algorithm. The method is based on the concept of conditional erosion [14]. Procedure assumes that the erosion is conducted as long as the size of the processed nucleus is large enough. Two masks for erosion operation are designed. They can be referred as fine and coarse erosion structuring elements. The coarse erosion tends to preserve the actual shape but reduces the size of clustered nuclei. This can make the nucleus to disappear because of huge reduction in the size. On the other hand, fine erosion mask is less likely to make the nucleus disappear, but it will lead to the loss of original shape. The erosion operation of the binary image $I$ by the structuring element $B$ is defined by:

$$I \ominus \check{B} = \{x \in \mathbb{R}^2 \mid (B + x) \subset I\},\qquad(1)$$

where $\check{B}$ is a reflection of set $B$. Conditional erosion is applied to binary image $BW$ obtained in the previous step of segmentation. Means of objects that have survived the conditional erosion become initial seeds used by the fast marching to segment individual nuclei Fig. 1(d).

Fast marching method is a special case of the level sets approach for monotonically advancing fronts. It was introduced by [10] and can be used to extract complex shapes from 2D and 3D images. In our work it is used to split the clustered nuclei. Algorithm starts with the initial front $\Gamma_0$. Next the front $\Gamma$ evaluates with speed $F(x, y)$ in the normal direction where $F$ is always either positive or negative. Front passes through a point $(x, y)$ at the time $T(x, y)$. Under this formulation the arrival time function $T(x, y)$ satisfies the Eikonal equation:

$$|\nabla T|F = 1. \tag{2}$$

In order to solve the equation, the gradient $|\nabla T|$ is estimated using upwind entropy-satisfying scheme. By limiting our considerations to two-dimensional grid, we must solve following quadratic equation:

$$1/F_{i,j}^2 = \max\left(\max(d_{i,j}^{-x}T, 0), -\min(d_{i,j}^{+x}T, 0)\right)^2$$
$$+ \max\left(\max(d_{i,j}^{-y}T, 0), -\min(d_{i,j}^{+y}T, 0)\right)^2, \tag{3}$$

where
$$d_{ij}^{\pm x}T = (T_{i\pm1,j} - T_{i,j})/h,$$
$$d_{ij}^{\pm y}T = (T_{i,j\pm1} - T_{i,j})/h, \tag{4}$$

and $h$ is the grid step. If the quadratic equation yields more than one solution, the greatest is chosen. The behavior of the front is driven by the speed function $F$. It must be designed in a way that the front stops exactly at the boundary of the nuclei. We decided to use speed function based on image local gradient:

$$F = e^{-\alpha|\nabla(H_\sigma * I)|}, \tag{5}$$

where $\alpha$ is a weighting factor, $I$ is the original image and $H_\sigma$ is a Gaussian smooth operator.

Standard fast marching is well suited to foreground-background segmentation. Nevertheless, our application must deal with multiple objects. It was realized by using multi-label fast marching [12]. Number of labels is determined by the number of nuclei detected by the conditional erosion. Each seed is associated with the unique label (segment). Even after applying fast marching procedure, we still can find spurious nuclei in the segmentation results. To exclude unwanted outliers, we decided to filter out objects with the area lower than 100 and bigger than 3000. Furthermore, we removed all objects with circularity ratio lower than 0.5 Fig. 1(e).

$$C_r = \frac{4\pi A}{p^2}, \tag{6}$$

where $C_r$ is a circularity ratio, $A$ is an object area and $p$ is an object perimeter. All these threshold values was chosen experimentally based on the knowledge about the shape and size of nuclei.

## 2.3   Feature Extraction

For each isolated nucleus 42 features are extracted. Then, for each image, the mean (M), median (D), standard deviation (T), kurtosis (K), skewness (S) and

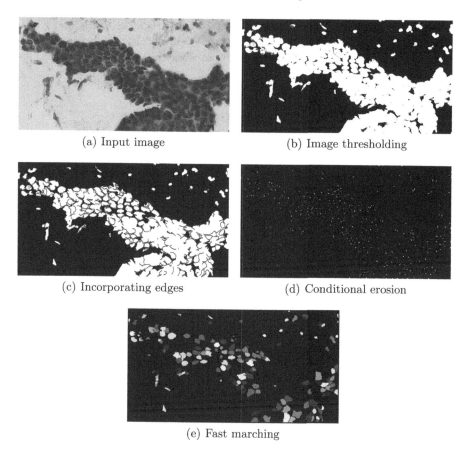

(a) Input image

(b) Image thresholding

(c) Incorporating edges

(d) Conditional erosion

(e) Fast marching

**Fig. 1.** Input image, final and intermediate results of the proposed segmentation procedure

interquartile range (I) are determined giving a total number of 252 features. The first group of features is related to the size and shape of the nuclei. This is represented by the following features: *Area* (A), *Perimeter* (P), *Eccentricity* (Ecc), *Major Axis Length* (MjAL), *Minor Axis Length* (MnAL), *Shape Factor* (SF), *Convex Deficiency* (CD), *Bending Energy* (BE).

The second group of features is related to the distribution of nuclei in the image. Healthy tissue usually form single-layered structures, while cancerous cells tends to break up which increases the probability of encountering separated nuclei. To express this relation, features representing the distance to centroid of all nuclei, and the distance to k-nearest nuclei are used: *Distance to Centroid of All Nuclei* (D2A), *Distance to c-Nearest Nuclei* (D2cNN).

The third group of features is related to the distribution of chromatin in the nuclei. This is represented with texture features based on gray-level co-occurrence matrix (GLCM) [2] and gray-level run-length matrix (GLRLM) [13], as well as the mean and variance of pixel values in each RGB channel.

Based on the statistics computed for the images, the same statistics are determined for cases (patients).

## 2.4   Classification

For classification, we use Naive Bayes model. Predictive accuracy was calculated with the leave-one-out procedure. There were 50 folds (the number of patients), and each fold consist of single patient (11 images). Two measures of the classification accuracy were defined:

- *patient accuracy* - the percentage ratio of successfully diagnosed cases (patients) to the total number of cases,
- *image accuracy* - the percentage ratio of successfully classified images to the total number of images,

To determine the image accuracy, all of the images were classified individually but with the restriction that the images belonging to the same patient were never at the same time in the training and testing set. Patient accuracy was determined by classifying cases described by features aggregated over 11 images. A suboptimal sets of input features were determined using sequential forward selection algorithm.

## 3   Experimental Results

In order to verify the effectiveness of the proposed nuclei segmentation approach, we applied this method to extract nuclei from 550 cytological images. These images came from cytological examinations performed for 50 patients. Every patient was described by 11 images (see Sect. 2.1 for details). Then, for each image, 252 features were extracted as in Sect. 2.3. Based on image features, aggregated statistics was calculated for every case (patient). Segmentation accuracy was measured in terms of predictive accuracy of Naive Bayes model. Taking into account the fact that testing set is quite small, we decided to estimate predictive accuracy using leave-one-out procedure. Results of this study are presented in the Table 1. Sequential forward selection was employed to choose suboptimal set of input features. The procedure choose 6 and 8 input features respectively for the patient accuracy and image accuracy. To see how the fusion of image thresholding with fast marching method improved the nuclei segmentation, we presented an illustrative example in Fig. 2. We can observe in Fig. 2(b) that image thresholding alone is not able to segment nuclei properly. In contrast, our approach is able to segment sufficient number of nuclei to prepare the diagnosis Fig. 2(a).

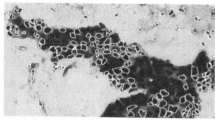
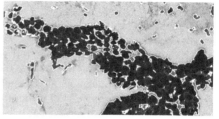

(a) Isodata image thresholding + fast marching

(b) Isodata image thresholding

**Fig. 2.** Results of nuclei segmentation

**Table 1.** Classification results (D, T, S, K and I after underscore are median, standard deviation, skewness, kurtosis and interquartile range respectively)

|  | Predictive accuracy (Naive Bayes) | Input features |
|---|---|---|
| Accuracy (patients) | 98 % | D2NNN_I, VR_S, GLCM_01_K, Ecc_S, SF_I, GLRLM_03_S |
| Accuracy (images) | 87.64 % | D2NNN_I, MnAL_S, P_T, MB_D, VR_T, MR_D, Ecc_T, VG_S |

# 4    Conclusions

Content of cytological images is highly complex and its analysis is difficult in an automated way. The crucial step of this analysis is nuclei segmentation. Generally such methods of image processing as image thresholding, edge detection or active contours are not able to extract nuclei with satisfactory quality. Thus, we can not build model able to classify cancerous nuclei. To tackle this problem we proposed to combine different image processing algorithms [1]. The effectiveness of the proposed approach was measured in terms of predictive accuracy of Naive-Bayes model. To ensure the reliability of the results, leave-one-out cross-validation was used to compute the predictive accuracy. Our study shows that proposed segmentation procedure allows to compute nuclei features that discriminate benign from malignant cases with predictive accuracy equal to 98 % for patients and 87.6 % for images. In the case of the patient classification, this means that only one case was wrongly classified. We observed that proposed segmentation method is not able to segment all nuclei, however the number of extracted nuclei is sufficient for diagnosis. We consider that further studies in this area will be necessary to test the effectiveness of the method when larger and richer database of cytological images will be available.

**Acknowledgments.** The research was supported by National Science Centre, Poland (2015/17/B/ST7/03704).

# References

1. Choraś, R., Andrysiak, T., Choraś, M.: Integrated color, texture and shape information for content-based image retrieval, pattern analysis and applications. Pattern Anal. Appl. **10**(4), 333–343 (2007)
2. Haralick, R., Shanmugam, K., Dinstein, I.: Textural features for image classification. IEEE Trans. Syst. Man Cybern. **3**(6), 610–621 (1973)
3. Jeleń, L., Fevens, T., Krzyżak, A.: Classification of breast cancer malignancy using cytological images of fine needle aspiration biopsies. Int. J. Appl. Math. Comp. Sci. **18**(1), 75–83 (2010)
4. Kowal, M., Filipczuk, P.: Nuclei segmentation for computer-aided diagnosis of breast cancer. Int. J. Appl. Math. Comp. Sci. **24**(1), 19–31 (2014)
5. Kowal, M., Filipczuk, P., Obuchowicz, A., Korbicz, J., Monczak, R.: Computer-aided diagnosis of breast cancer based on fine needle biopsy microscopic images. Comput. Biol. Med. **43**(10), 1563–1572 (2013)
6. Kruk, M., Osowski, S., Markiewicz, T., Slodkowska, J., Koktysz, R., Kozlowski, W., Swiderski, B.: Computer approach to recognition of fuhrman grade of cells in clear-cell renal cell carcinoma. Anal. Quant. Cytopathol. Histpathol. **36**(3), 147–160 (2014)
7. Plissiti, M., Nikou, C., Charchanti, A.: Combining shape, texture and intensity features for cell nuclei extraction in pap smear images. Pattern Recogn. Lett. **32**(6), 838–853 (2011)
8. Ridler, T.W., Calvard, S.: Picture thresholding using an iterative selection method. IEEE Trans. Syst. Man Cybern. **SMC–8**(8), 630–632 (1978)
9. Savchenko, A.V., Belova, N.S.: Statistical testing of segment homogeneity in classification of piecewise-regular objects. Int. J. Appl. Math. Comp. Sci. **25**(4), 915–925 (2015)
10. Sethian, J.: A fast marching level set method for monotonically advancing fronts. Proc. Natl. Acad. Sci. **93**(4), 1591–1595 (1996)
11. Sezgin, M., Sankur, B.: Survey over image thresholding techniques and quantitative performance evaluation. J. Electr. Imaging **13**(1), 146–168 (2004)
12. Steć, P.: Segmentation of Colour Video Sequences using the Fast Marching Method. University of Zielona Góra Press, Zielona Góra (2005)
13. Tang, X.: Texture information in run-length matrices. IEEE Trans. Image Process. **7**(11), 1602–1609 (1998)
14. Yang, X., Li, H., Zhou, X.: Nuclei segmentation using marker-controlled watershed, tracking using mean-shift, and kalman filter in time-lapse microscopy. IEEE Trans. Circ. Syst. I Regul. Pap. **53**(11), 2405–2414 (2006)

# Multiclass AdaBoost Classifier Parameter Adaptation for Pattern Recognition

Jerzy Dembski$^{(\boxtimes)}$

Faculty of Electronics, Telecommunications and Informatics,
Gdansk University of Technology, ul. Narutowicza 11/12, 80-952 Gdansk, Poland
`dembski@ue.eti.pg.gda.pl`

**Abstract.** The article presents the problem of parameter value selection of the multiclass "one against all" approach of an AdaBoost algorithm in tasks of object recognition based on two-dimensional graphical images. AdaBoost classifier with Haar features is still used in mobile devices due to the processing speed in contrast to other methods like deep learning or SVM but its main drawback is the need to assembly the results of binary two-class classifiers in recognition problems. In this paper an original method of selecting the parameter values of the assembling algorithm using many similar face recognition tasks is proposed. The parameter optimization is done by checking all possible vectors of parameter values. The recognition results with optimized parameter values is 10 % better in 8-class face database famous48 (http://eti.pg.edu. pl/documents/176468/27493127/famous48.zip) tasks than using random heuristic which can be represented by the average of all possible vectors of parameter values.

**Keywords:** Face recognition · Multiclass AdaBoost classifier · Haar features

## 1 Introduction

One of the basic criteria for the assessment of pattern recognition systems beyond the classification accuracy of test examples (generalisation) is the recognition speed. It is crucial in some applications, such as object tracking in real time or gesture control interfaces. In the tasks of object recognition in graphical images two kinds of methods are currently used: specialized methods which utilise knowledge of recognized specific types of objects and more universal methods. In the case of face recognition a typical specialised method designates the individual characteristic points (centers of eyes, eyebrows, the center of mouth, etc.) and then calculates and compares the parameters of the Gabor filters in these characteristic points using simple classifier such as a nearest neighbour classifier.

Examples of more general methods which give very high generalisation are support vector machine (SVM) with a histogram of oriented gradient features (HOG) [1] and convolutional neural networks (CNN) [6]. Both methods give very

© Springer International Publishing AG 2017
R.S. Choraś (ed.), *Image Processing and Communications Challenges 8*,
Advances in Intelligent Systems and Computing 525, DOI 10.1007/978-3-319-47274-4_24

small generalisation error but are not enough quick for using in real time applications especially in mobile devices. The cascaded version of AdaBoost using Haar features [8] or other features like Census [5] or HOG [1] in binary version [2] is an example of the method with worse generalisation but quick enough. In the case of a sequence of images, worse generalisation can be compensated by merging particular images recognition results using HMM.

The main drawback of AdaBoost classifier in recognition tasks is that this is a two-class (binary) classifier and in $K$-class cases the sophisticated multiclass direct version of AdaBoost algorithm or simple "one vs all" approach is needed. The main disadvantage of direct methods like AdaBoost.M1 [4] and SAMME [9] is the need to use multiclass "strong" weak classifiers which contradict the weak classifier idea and may reduce generalisation. In AdaBoost.M2 [4] and in AdaBoost.OC [7] algorithms the strong classifiers are assembled with binary weak classifiers. All direct methods do not allow to use many similar classification tasks to optimise their performance in easy way which is the main contribution of this work.

Due to above disadvantages of direct multiclass AdaBoost algorithms, in this work the simple approach "one vs all" was taken into consideration.

## 2    The Binary AdaBoost Classifier with Haar-Like Features for Object Detection

The main idea of boosting [4] is to improve overall strong classifier by adding subsequent weak classifiers which are as simple as possible. The vector of training examples weights is used to pay more attention to wrongly classified examples in the previous steps.

### 2.1    Strong AdaBoost Classifier Training

At each step of the while loop in Table 1 the best weak classifier due to weighted error from a great quantity of possible weak classifiers is chosen. At the last step of while loop the best threshold value $\Theta$ of a strong classifier is calculated using validation or mixed set of examples. The strong classifier training process stops when the false positive error $f <= f_{\max}$ providing that true positive rate is over assumed value $d >= d_{\min}$. The strong classification function $H(\mathbf{x})$ consisted with $T$ weak classifiers as a result of AdaBoost training process is described by the formula:

$$H(\mathbf{x}) = \begin{cases} 1 & \text{if} \quad S = \sum_{t=1}^{T} \alpha_t h_t(\mathbf{x}) \geq \Theta \\ 0 & \text{otherwise,} \end{cases} \tag{1}$$

where $\alpha_t = \log(1 - \epsilon_t) - \log \epsilon_t$ is a weight of $t$-th weak classifier, $h_t(\mathbf{x})$ is $t$-th weak classifier output value, $\mathbf{x}$ is an input image pixel intensity vector, $\Theta$ is a threshold.

**Table 1.** AdaBoost strong classifier (single layer in a cascade) learning algorithm

---

**for** each training example $\{(\mathbf{x}_1, c_1), (\mathbf{x}_2, c_2) \dots (\mathbf{x}_N, c_N)\}$

> initialize weights $w_i = 1/N_c$, where $N_c$ is a number of examples which belong to the same class $c$ as $i$-th example

**end for**

**while**   false positive error $f > f_{\max}$ **or** true positive rate $d < d_{\min}$

1. normalise weights: $w_i \leftarrow \frac{w_i}{\sum_{j=1}^{N} w_j}$
2. select the weak classifier $h_t(\mathbf{x})$, which minimises the weighted classification error: $\epsilon_t = \min_j \sum_{i=1}^{N} w_i |h_j(\mathbf{x}_i) - c_i|$
3. decrease the weights of properly classified examples: $w_i \leftarrow w_i \frac{\epsilon_t}{1-\epsilon_t}$
4. find the best threshold value $\Theta$ for strong classifier with minimum false positive error if a false negative error is under expected value for a node (cascade layer) $d >= d_{\min}$

**end while**

---

## 2.2   The Cascaded Version of AdaBoost Classifier for Object Detection

The main idea of cascaded version [8] of AdaBoost algorithm (Fig. 1) is to reduce time complexity thanks to early rejection images which don't pass early cascade layers. Each layer is trained to reject not so big part of negative images, for instance 50 % which results in use only a few arithmetical operations per layer. It is especially important in the case of unbalanced datasets e.g. when there are many times more negative examples than positive ones. During a cascade learning process each layer in Fig. 1 is trained using strong AdaBoost classifier training algorithm described in Sect. 2.1 using examples which are not rejected in previous layers.

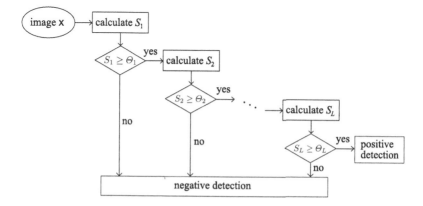

**Fig. 1.** The AdaBoost cascade

# 3   The Binary Classifiers Assembling Algorithm

The first training stage in the simple $K$-class AdaBoost in "one vs all" fashion training process is to train $K$ binary classifiers each for distinguishing $k$-th class objects from all other objects. This purpose is achieved by $K$ independent cascade training processes described in Sect. 2.2 with specific labeling of images: in each $k$-th process label 1 (positive) is assigned for images from class $k$ and label 0 (negative) for images from other classes.

In the first classification step the input 2D image is presented to each of early prepared $K$ binary classifiers and after summation it can lead to one of three possible cases:

1. $K_{pos} = 1$ – exactly one binary classifier classifies an image as positive one,
2. $K_{pos} > 1$ – more than one classifies an image as positive one,
3. $K_{pos} = 0$ – all binary classifiers classify an image as negative one,

where $K_{pos}$ is a number of binary classifiers which classify an image as positive example. In the case (1) the decision of multiclass classifier is obvious: the class of image is determined by the positively returned binary classifier. In the cases (2) and (3) the usage of additional information is necessary. For instance the couples of values $(S_i, \Theta_i)$ for each cascade layer $i = 1 \ldots L$ can be used in many configurations as differences or quotients between $S_i$ and $\Theta_i$. It is not known which is better especially in connection with methods of combining results from particular cascade layers to take into account which layers are more important. These problems suggest a conclusion that binary classifiers assembling algorithm in second training stage can be optimised based on its parameter values adaptation. As the minimization of generalisation error is a purpose of optimisation, the reliable parameter values can be obtained by using many recognition tasks as "training examples" at the second stage.

## 3.1   The Assembling Algorithm Parametrization

In the most general method we can treat all the accessible cascade parameters as $\Theta_i, S_i, i = 1 \ldots L$ from all binary classifiers as input variables of unknown function that assigns the class number for an input image. This function can then be approximated by a multilayer artificial neural network or symbolic expression tree so that the optimisation can be done by simulated annealing, genetic algorithm or genetic programming. This approach has some drawbacks such as the variability of the number of input values due to the different number of layers or too great complexity of such optimisation tasks which would have required a large amount of training data and a very long training.

A simpler solution which is proposed in this work consists of assuming a simple assembling algorithm with a constant quantity of heuristically chosen binary parameters which can be adapted using enumerated searching. In the $K_{pos} > 1$ cases when more than one binary classifier returns a positive answer 5 important parameters using vectors of 6 binary values were chosen which was described in Table 2. For the $K_{pos} = 0$ cases a vector of 7 binary values were chosen (Table 3).

**Table 2.** Parametrization for $K_{pos} > 1$ cases

| Bit number | Parameter interpretation | Binary values |
|---|---|---|
| 1 | Relation type | 0 - difference: $S - \Theta$<br>1 - quotient: $S/\Theta$ |
| 2, 3 | Type of layers weighting | 0, 0 - exponential distribution<br>0, 1 - linear distribution<br>1, 0 - extremal value<br>1, 1 - constant weights |
| 4 | Direction of weight changes | 0 - decreasing or minimum<br>1 - increasing or maximum |
| 5 | Normalisation | 0 - values divided by sum of weights<br>1 - direct values |
| 6 | Subset of detectors | 0 - all $L$ detectors<br>1 - only positive detectors |

**Table 3.** Parametrization for $K_{pos} = 0$ cases

| Bit number | Parameter interpretation | Binary values |
|---|---|---|
| 1–5 | The same as 1-5 bits for $K_{pos} > 1$ cases | |
| 6 | Criterion type | 0 - $(S - \Theta)$ or $(S/\Theta)$<br>1 - length of layer sequence with positive detections |
| 7 | Number of layers taken into account | 0 - all layers<br>1 - only layers with positive detections |

## 3.2   The Framework of Optimisation Procedure

Each vector of parameter values can be treated as a sequence of bits: $(p_1, p_2, \ldots, p_6)$ for $K_{pos} > 1$ cases and $(q_1, q_2, \ldots, q_7)$ for $K_{pos} = 1$ cases. There are $2^6 = 64$ possible vectors for $K_{pos} > 1$ cases and 128 for $K_{pos} = 0$. Taking into account that both types of cases are disjunctive, each case can be separately optimised using below framework:

1. choose random binary values for $K_{pos} = 0$, $\mathbf{q^0} = (1, 1, 1, 0, 1, 1, 1)$,
2. find optimal values for training tasks $\mathbf{p} = (p_1, p_2, \ldots, p_6)$ for $K_{pos} > 1$,
3. find optimal values $\mathbf{q} = (q_1, q_2, \ldots, q_7)$ for $K_{pos} = 0$.

Earlier $S$ values calculation for all cascade layers and all example images allow all possible value sequences to check one by one with low time complexity.

## 3.3   Description of Assembling Algorithm Parameters

The first parameter represents the relation type $r$. If $p_1 = 1$ it is a quotient of a sum of weighted weak classifiers and threshold value $r_i = S_i/\Theta_i, i = 1 \ldots L$, where $L$ is the number of layers, $S = \sum_{t=1}^{T} \alpha_t h_t(\mathbf{x})$ is a weighted sum of weak classifiers decisions in a $i$-th layer, else if $p_1 = 0$, $r_i = S_i - \Theta_i$. In both cases the strong classifier decision $H_i(\mathbf{x})$ is more reliable as the $r_i$ value is larger.

This relation has to be calculated for each cascade layer. The importance of particular layers is expressed by the second parameter represented by a couple of bits $p_2, p_3$ which describes the weights in weighted sums of chosen cascade layer relations. If $(p_2 = 0, p_3 = 1)$, the weights grow or decline linearly depended on parameter $p_4$: $w_i = i$ when $p_4 = 1$ or $w_i = L - i$ when $p_4 = 0$, where $i = 1 \ldots L$. In the case $(p_2 = 0, p_3 = 1)$, $w_i = i$ when $p_4 = 1$ or $w_i = L - i$ when $p_4 = 0$. In the case $(p_2 = 1, p_3 = 1)$, the importance for all layers are equal: $w_i = 1$. The class of image is described by the number of a binary classifier with the greatest $S$ value:

$$C(\mathbf{x}) = \underset{k}{\operatorname{argmax}} \frac{\sum_{i=1}^{L_k} w_{ik} r_{ik}}{\sum_{i=1}^{L_k} w_{ik}}, \tag{2}$$

where $k$ is a number of a binary classifier. If $p_5 = 1$ the weighted sum is not normalised e.g. in Eq. 2 the denominator is empty. The normalisation described by $p_5$ is important according to the unrestricted number of layers $L$. The case $(p_2 = 1, p_3 = 0)$ is an exception when the minimum value is taken into account $\min_i(r_i), i = 1 \ldots L$ if $p_4 = 0$ or $\max_i(r_i)$ if $p_4 = 1$.

The first 5 binary parameters in the case $K_{pos} = 0$ are analogical to $K_{pos} > 1$ case. If the additional parameter $q_6 = 1$, the number of layers in a sequence of positive detections are a criterion inside argmax(.) expression in Eq. 2 instead of weighted sum of $r$. The parameter $q_7$ is used to indicate which layers are taken to weighted sum evaluation in Eq. 2: if $q_7 = 0$, all layers and if $q_7 = 1$, only layers with positive detections.

### 3.4   Data Preparation and Binary Classifier Training Stage

The face recognition tasks with its own face database famous48 are used for experiments of assembling algorithm parameter optimisation. The database contains faces of 48 famous people (classes). Each class is represented by average 150 gray scale 24×24 images, summing up to 6835 example images. Each image has constant mean and variation of pixel intensities. The face placement inside an image is also normalised. The centers of eyes are always placed at the same points of image on the horizontal line, a center of mouth on the vertical line which bisects an image.

The set of 48 personal classes was divided into 6 subsets each containing 8 classes. In each subset of classes the set of all images from these classes was divided into subsets of training and test examples in 4 different ways. The test examples contribution is about 25 % of all examples. In such a way $6 * 4 = 24$ 8-class face recognition tasks were generated. In the first training stage 8 binary classifiers for each task were trained by an AdaBoost algorithm using Haar-like features and modular detection system [3].

## 4   Experiments and Optimization Results

Table 4 presents optimisation results for 5 experiments with different recognition tasks for parameter optimisation (training tasks) and for test. Each number in the second or third column denotes the subset of people e.g. classes $\{1, 2, \ldots, 8\}$ for number 1, classes $\{9, 10, \ldots, 16\}$ for number 2 and so on. The quantity of tasks in the table should be multiplied by 4 because of 4 ways of training/test examples division ways.

**Table 4.** The assembling parameter optimisation results

| Exp. No. | Training tasks | Test tasks | Values of chosen parameters | Test example classification average accuracy in test tasks | | | |
|---|---|---|---|---|---|---|---|
| | | | | Chosen params | Mean | Max | Min |
| 1 | 1, 2, 3, 4 | 5, 6 | $\mathbf{p} = (1, 0, 0, 0, 1, 1)$ $\mathbf{q} = (1, 0, 1, 0, 0, 0, 0)$ | **0.885** | 0.797 | 0.885 | 0.676 |
| 2 | 3, 4, 5, 6 | 1, 2 | $\mathbf{p} = (1, 0, 0, 0, 1, 0)$ $\mathbf{q} = (1, 0, 1, 0, 0, 0, 0)$ | **0.825** | 0.721 | 0.830 | 0.579 |
| 3 | 1, 2, 5, 6 | 3, 4 | $\mathbf{p} = (1, 0, 1, 0, 0, 0)$ $\mathbf{q} = (1, 0, 1, 0, 0, 0, 0)$ | **0.811** | 0.705 | 0.816 | 0.563 |
| 4 | 1, 3, 5 | 2, 4, 6 | $\mathbf{p} = (1, 0, 0, 0, 1, 1)$ $\mathbf{q} = (1, 0, 1, 0, 0, 0, 0)$ | **0.857** | 0.761 | 0.857 | 0.630 |
| 5 | 2, 4, 6 | 1, 3, 5 | $\mathbf{p} = (1, 0, 0, 0, 1, 0)$ $\mathbf{q} = (1, 0, 1, 0, 0, 0, 0)$ | **0.827** | 0.721 | 0.828 | 0.583 |

In each optimisation experiment the best vector of parameter values for $K_{pos} > 1$ cases e.g. $(1, 0, 0, 0, 1, 1)$ in first experiments is used to evaluate results of $K_{pos} = 0$ cases parameter optimisation according to the framework in Sect. 3.2. The most significant results of optimisation are written by bold fonts and describe the average classification accuracy of test examples in test tasks e.g. tasks enumerated in third column for the chosen (best for training tasks) parameter value sequence. In each experiment an improvement in comparison to mean value averaged for all possible vectors of the parameter values (in the next column) was indicated. This mean value can be representative for results obtained by a random heuristic assembling algorithm.

## 5   Conclusions

The AdaBoost multiclass assembling algorithm parameter optimisation method proposed in this work allows us to obtain about 10 % lower average generalisation error in test (unknown) recognition tasks than random heuristic (mean value) as was showed in Table 4. The novelty in comparison to other methods is due to usage of many similar multiclass recognition tasks from the same domain to obtain high reliability of estimating solution. In this work the face recognition domain was used to find the best solution.

In future research the following issues should be investigated:

- checking whether obtained results are valid also in other object type recognition or even in any classification tasks,
- direct multiclass versions like AdaBoost.M2, OC, ECC optimisation using many similar tasks,
- checking whether the assembling algorithm optimisation becomes more important in decreasing amount of training examples.

All above issues need to retrain binary classifiers which is very time consuming.

# References

1. Dalal, N., Triggs, B.: Histograms of oriented gradients for human detection. In: IEEE Computer Society Conference on Computer Vision and Pattern Recognition (CVPR 2005), pp. 886–893 (2005)
2. Dembski, J.: Multiscaled hybrid features generation for adaboost object detection. J. Med. Informatics Technol. **24**(2015), 75–82 (2015)
3. Dembski, J., Smiatacz, M.: Modular machine learning system for training object detection algorithms on a supercomputer. In: Advances in System Science, pp. 353–361 (2010)
4. Freund, Y., Schapire, R.E.: Experiments with a new boosting algorithm. In: Proceedings of the Thirteenth International Conference on Machine Learning, pp. 148–156 (1996)
5. Küblbeck, C., Ernst, A.: Face detection and tracking in video sequences using the modified census transformation. Image Vis. Comput. **24**(6), 564–572 (2006)
6. LeCun, Y., Boser, B., Denker, J.S., Henderson, D., Howard, R.E., Hubbard, W., Jackel, L.D.: Backpropagation applied to handwritten zip code recognition. Neural Comput. **1**(4), 541–551 (1989)
7. Schapire, R.E.: Using output codes to boost multiclass learning problems. In: Proceedings of the Fourteenth International Conference on Machine Learning (ICML 1997), pp. 313–321 (1997)
8. Viola, P., Jones, M.: Robust real-time face detection. Int. J. Comput. Vision **57**(2), 137–154 (2004)
9. Zhu, J., Rosset, S., Zou, H., Hastie, T.: Multi-class adaboost. Stat. Interface **2**, 349–366 (2009)

# Communications

# CIPRNet Training Lecture: Hybrid Simulation of Distributed Large-Scale Critical Infrastructures

Massimo Ficco[✉]

Department of Industrial and Information Engineering,
Second University of Naples (SUN), Via Roma 29, 81031 Aversa, Italy
ficco@unina.it

**Abstract.** Modern critical infrastructures represent the pivotal assets upon which the current society greatly relies to support welfare, economy, and quality of life. Nowadays, the trend is to re-organize these infrastructures by applying a System of Systems concept, where the sparse islands are progressively interconnected by means of proper middleware solutions through local or wide-area networks. The huge complexity of such systems makes the integration task among components extremely challenging. Indeed, it may introduce unexpected system behaviors, mainly affecting dependability and performance, that usually become evident only during systems operations and, in particular, in presence of stress or unexpected conditions. Additionally, as they cannot be detected earlier, these problems require complex on-site operations resulting in increased maintenance costs and overspending in terms of personnel resources. A promising way to cope with these new complex systems and to reduce maintenance costs, is to reproduce such distributed systems locally, and let them run prior to the actual execution on-site, in order to get knowledge about their real behavior and define mitigation means and improvement actions. On the other hand, the evaluation of this systems requires sophisticated modeling, simulation, and experimentation infrastructure, which needs the integration of existing simulation environments, real sub-systems, and experimental platforms, which have to interact in a coordinated way. Therefore, hybrid and distributed simulation strategies, supported by novel technologies for resources virtualization and working environment reproduction, represent the most promising way to define the needed strategies to actually support such complex paradigms [1,2].

**Keywords:** Cloud federation · Mission-critical applications · Cloud elasticity · Security

**Acknowledgment.** This lecture is supported by CIPRNet Network of Excellence in Critical Infrastructure Protection, available at https://www.ciprnet.eu/summary.html.

R.S. Choraś (ed.), *Image Processing and Communications Challenges 8*,
Advances in Intelligent Systems and Computing 525, DOI 10.1007/978-3-319-47274-4_25

# References

1. Ficco, M., Avolio, G., Palmieri, F., Castiglione, A.: An HLA-based framework for simulation of large-scale critical systems. Concurrency Comput. Pract. Experience **28**(2), 400–419 (2016)
2. Ficco, M., Di Martino, B., Pietrantuono, R., Russo, S.: Optimized task allocation on private cloud for hybrid simulation of large-scale critical systems. Fut. Gener. Comput. Syst. (2016)

# Latin Multiplication in Telemetry Hot Potato Wireless Sensor Networks Analysis

Ireneusz Olszewski[✉]

Faculty of Telecommunications, Informatics and Electrical Engineering,
Institute of Telecommunications, UTP University of Science and Technology,
S. Kaliskiego 7, 85-789 Bydgoszcz, Poland
ireneusz.olszewski@utp.edu.pl

**Abstract.** This paper proposes the use of the Latin Multiplication for the WSNs (Wireless Sensor Networks), in which Hot-Potato protocol (H-P) is used. It is shown that making small modifications in the protocol increases the probability of reaching a destination node by the packet injected by the source node. The same rule was observed in the direction from thin node to the sink. The probability is determined for the assumed number of hops. The results of this research show us that from practical point of view the most useful solution is to direct the packet along a directed route, which does not consist of cycles with the length equals to two.

**Keywords:** WSN (Wireless Sensor Networks) · Hot-potato protocol · Path · Directed route

## 1 Introduction

Wireless Sensor Networks consist of many inexpensive nodes, which are equipped with sensors and a radio transceivers which enable communication with other nodes in the network. Sometimes nodes are distributed on a quite large area. What is most characteristic for these nodes, is their low power energy consumption, because in most cases they are supplied from batteries and are expected to work for a long time [5]. WSNs have a wide range of applications, including: military, seismic applications, access control and surveillance applications and are also used in tracking applications for certain animal species. They can also act as "last mile" networks in advanced measurement AMI systems, which are used for automatic meter reading [9], e.g. electricity, gas consumption, etc. In case of the WSN node installed as electricity meters or used in smart lighting systems [7] the energy issues do not occur. In WSNs one of the node acts as the acquisition node. The task of this node is to collect, store and process information from other nodes in the network, while the remaining nodes perform measurement functions, or additionally perform as the transit nodes, if the multi-hop technique was implemented. On the one hand, the important advantage of WSN networks is the high rate of data acquisition and distribution, which is very

© Springer International Publishing AG 2017
R.S. Choraś (ed.), *Image Processing and Communications Challenges 8*,
Advances in Intelligent Systems and Computing 525, DOI 10.1007/978-3-319-47274-4_26

important from the point of view of the whole system [2]. On the other hand, the lifetime of the nodes depends on the power source capacity. Regardless of the ability to recover energy from the environment by using various kinds of transducers, e.g. photovoltaic cells, it is crucial to minimize the power consumption of the network nodes. Therefore, the basic assumption of the proposed communication protocols was to solve the problem of long-term nodes power supply [1]. However, in this work the WSN is considered in the context of the telemetric system for monitoring the electric energy consumption.

## 2    WSNs Communication Protocols

Due to the fact that in this work only WSNs for monitoring the electric energy consumption are considered, the old problem of WSN node power supply does not occur. Therefore, it is assumed that only simple, energy greedy communication protocols will be considered. These protocols are mostly based on the "multi-path" and "multi-hop" techniques [1,9]. This assumption allows to realize transmission with high reliability, because using "multi-path" technique, the packets are transmitted via many different paths and it also allows to reduce the memory capacity of the nodes. This last advantage is very important because of the fact that the nodes of the systems for remote meter reading use their large area of the memory encryption algorithms implementation. In addition, an increasing part of RAM is used for the implementation of the transceiver buffers, due to the need of sending longer and longer packets via WSN. Despite these advantages the protocols based on "multi-hop" and "multipath" have their drawbacks, among which the most frequently mentioned is high emission as result of the need of setting the paths. Due to this fact the authors of [4], for the first time proposed the use of Hot Potato (H-P) protocol for WSNs dedicated for the monitoring of the electricity consumption. The general idea of H-P protocol is based on the immediate packet sending by a monitored node to one of the adjacent nodes, which then forwards the packet to the next node etc. until the packet will be received by the acquisition node. Because the packet is immediately transmitted by intermediate node to the next node, packet buffering is not required. Only the addresses of neighboring nodes must be kept in their memory. The existence of only a single packet in the whole network at any given time excludes the possibility of a collision. It is why H-P protocol is classified as low emission protocol, unlike the "multi-path" protocol. As is apparent from the description, the path that a packet travels from a monitored node to the acquisition node (as well as in the opposite direction) can be described by directed route, which means that the nodes and arcs may even be repeated multiple times. H-P protocol was described for the first time in the mid-sixties, however, the widespread use occurred only in the second half of the nineties, when researchers realized that it is better to send packets over high speed fiber links via even more nodes than buffer these packets until the release of the optimal direction [3,11]. In this work, it is proposed to use the Latin Multiplication for the Hot Potato protocol employed in WSNs, which is used AMR. Latin Multiplication provides a simple way to make various modifications to the Hot Potato protocol.

The work is organized as follows: the method, with the usage of the Latin Multiplication, of H-P WSN network analysis is shown in the 3rd section; in the 4th section results for the exemplary networks, which is analyzed in the 3rd section, are presented; the 5th section contains a summary and conclusions.

## 3   WSN Analysis

As it was previously explained, the subject of this analysis is the Wireless Sensor Network used for the consumption profiles reading of the electricity counters. The most frequently used frequency bands for the implementation of such networks are the ISM (Industrial Scientific and Medical) bands i.e. 433 MHz, 866 MHz 2.4 GHz. In Fig. 1(a) the network under consideration is shown. This network consist of seven nodes signed from $a$ to $g$, a transmission range of each the nodes is also presented in this figure. In turn, the topological structure of the network, where each edge of the graph is a bi-directional link. Such a network can be described using the graph, where nodes of the network are represented by the nodes of the graph, while the one-direction links are represented by the arcs of the graph. In this case, the network will be described in a binary matrix of linear transformation $P$ with $n$ rows and $n$ columns, in which $p_{ij}$ is equal to 1, when there is an arc between nodes $i$ and $j$, otherwise it is equal to 0.

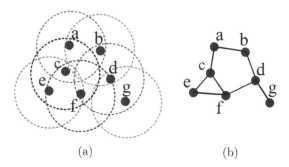

(a)                                    (b)

**Fig. 1.** Analyzed sensor network (a) nodes distribution together with their radio range (b) topological structure of the network

Matrix of linear transformation for the graph presented in Fig. 1(b) is as follow:

$$P = \begin{bmatrix} 0 & 1 & 1 & 0 & 0 & 0 & 0 \\ 1 & 0 & 0 & 1 & 0 & 0 & 0 \\ 1 & 0 & 0 & 0 & 1 & 1 & 0 \\ 0 & 1 & 0 & 0 & 0 & 1 & 1 \\ 0 & 0 & 1 & 0 & 0 & 1 & 0 \\ 0 & 0 & 1 & 1 & 1 & 0 & 0 \\ 0 & 0 & 0 & 1 & 0 & 0 & 0 \end{bmatrix}$$

## 3.1   Hot Potato Protocol

It has been already explained in previous unit, that in H-P protocol, the packet that was sent by a monitored node to the acquisition node travels along the directed route of some length. To make the analysis of such a network, it is necessary to define the number of directed routes between all the pairs of the nodes. It is known, from graph theory, that element $p_{ij}^{(k)}$ a $k^{\text{th}}$ powered matrix $P$ equals to the number of the $k$-length directed routes between the nodes $i$ and $j$ [10]. Below, the second, the third and the fourth power of the matrix $P$ is presented, respectively.

$$P^{(2)} = \begin{bmatrix} 2 & 0 & 0 & 1 & 1 & 1 & 0 \\ 0 & 2 & 1 & 0 & 0 & 1 & 1 \\ 0 & 1 & 3 & 1 & 1 & 1 & 0 \\ 1 & 0 & 1 & 3 & 1 & 0 & 0 \\ 1 & 0 & 1 & 1 & 2 & 1 & 0 \\ 1 & 1 & 1 & 0 & 1 & 3 & 1 \\ 0 & 1 & 0 & 0 & 0 & 1 & 1 \end{bmatrix} \quad P^{(3)} = \begin{bmatrix} 0 & 3 & 4 & 1 & 1 & 2 & 1 \\ 3 & 0 & 1 & 4 & 2 & 1 & 0 \\ 4 & 1 & 2 & 2 & 4 & 5 & 1 \\ 1 & 4 & 2 & 0 & 1 & 5 & 3 \\ 1 & 2 & 4 & 1 & 2 & 4 & 1 \\ 2 & 1 & 5 & 5 & 4 & 2 & 0 \\ 1 & 0 & 1 & 3 & 1 & 0 & 0 \end{bmatrix} \quad P^{(4)} = \begin{bmatrix} 7 & 1 & 3 & 6 & 6 & 6 & 1 \\ 1 & 7 & 6 & 1 & 2 & 7 & 4 \\ 3 & 6 & 13 & 7 & 7 & 8 & 2 \\ 6 & 1 & 7 & 12 & 7 & 3 & 0 \\ 6 & 2 & 7 & 7 & 8 & 7 & 1 \\ 6 & 7 & 8 & 3 & 7 & 14 & 5 \\ 1 & 4 & 2 & 0 & 1 & 5 & 3 \end{bmatrix}$$

Based on the obtained matrixes, it is possible to determine numbers of directed routes between the acquisition node and any other node in the network. It should be noted, that as the acquisition node was assumed the root of the graph tree, from the Fig. 1(b) i.e. $d$ node [4]. Due to the fact that the network has only seven nodes, a proper choice of the acquisition node will not have much significance. However, in the case of several tens or more nodes, this choice will affect the time of reading of all the counters. Below, Table 1 shows the number of directed routes between the $d$ node and other nodes in the network, which have the length of $k = 1, 2, ..., 6$, respectively.

**Table 1.** The number of directed routes of $k$ length from $d$ to other nodes

| Destination node | | | | | | | Sum of paths |
|---|---|---|---|---|---|---|---|
| $k$ | a | b | c | d | e | f | g |
| 1 | 0 | 1 | 0 | 0 | 0 | 1 | 1 | 3 |
| 2 | 1 | 0 | 1 | 3 | 1 | 0 | 0 | 6 |
| 3 | 1 | 4 | 2 | 0 | 1 | 5 | 3 | 16 |
| 4 | 6 | 1 | 7 | 12 | 7 | 3 | 0 | 36 |
| 5 | 8 | 18 | 16 | 4 | 10 | 26 | 12 | 94 |
| 6 | 34 | 12 | 44 | 56 | 42 | 30 | 4 | 222 |

## 3.2   First Modification of the Hot Potato Protocol

There is another method which allows to determine the number of straight paths along which the packet is transmitted from the source node to the destination at a given number of hops based on the Latin Multiplication [6]. It should be

|   | a | b | c | d | e | f | g |
|---|---|---|---|---|---|---|---|
| a | 0 | ab | ac | 0 | 0 | 0 | 0 |
| b | ba | 0 | 0 | bd | 0 | 0 | 0 |
| c | ca | 0 | 0 | 0 | ce | cf | 0 |
| d | 0 | db | 0 | 0 | 0 | df | dg |
| e | 0 | 0 | ec | 0 | 0 | ef | 0 |
| f | 0 | 0 | fc | fd | fe | 0 | 0 |
| g | 0 | 0 | 0 | gd | 0 | 0 | 0 |

**Fig. 2.** $M^{(1)}$ matrix

|   | a | b | c | d | e | f | g |
|---|---|---|---|---|---|---|---|
| a | 0 | b | c | 0 | 0 | 0 | 0 |
| b | a | 0 | 0 | d | 0 | 0 | 0 |
| c | a | 0 | 0 | 0 | e | f | 0 |
| d | 0 | b | 0 | 0 | 0 | f | g |
| e | 0 | 0 | c | 0 | 0 | f | 0 |
| f | 0 | 0 | c | d | e | 0 | 0 |
| g | 0 | 0 | 0 | d | 0 | 0 | 0 |

**Fig. 3.** $\tilde{M}^{(1)}$ matrix

noted that the straight path is described as a series of consecutive vertexes such that each of them is not repeated. Latin Multiplication is a type of matrix multiplication, which allows to determine all elementary paths, determined by the number of arcs of length 1, 2..., till Hamiltonian path, which has the length of $n-1$. This method is shown for the network of Fig. 1(b). On the basis of e.g. the neighborhood matrix or binary matrix of linear transformation, the Latin Multiplication - $M^{(1)}$ is created in the following way: if $(i,j)$-element of the neighborhood matrix equals to 1 then the $(i,j)$-element of $M^{(1)}$ matrix takes the value of $i$, $j$. Applying assumed symbology, $M^{(1)}$ matrix, for the graph of Fig. 1(b), is shown in Fig. 2.

It should be noted that the matrix $M^{(1)}$ defines all the possible paths of length equal to 1. Next, from this matrix, the next $\tilde{M}^{(1)}$ matrix is formed by removing the first letter of each element of the matrix. Obtained $\tilde{M}^{(1)}$ matrix is shown in Fig. 3.

|   | a | b | c | d | e | f | g |
|---|---|---|---|---|---|---|---|
| a | 0 | 0 | 0 | abd | ace | acf | 0 |
| b | 0 | bac | 0 | 0 | 0 | bdf | bdg |
| c | 0 | cab | 0 | cfd | cfe | cef | 0 |
| d | dba | 0 | dcf | 0 | dfe | 0 | 0 |
| e | eca | 0 | efc | efd | 0 | ecf | 0 |
| f | fca | fdb | fec | 0 | fce | 0 | fdg |
| g | 0 | gdb | 0 | 0 | 0 | gdf | 0 |

**Fig. 4.** $M^{(2)}$ matrix

Here, by matrix multiplication (line by column) the $M^{(2)} = M^{(1)}\tilde{M}^{(1)}$ is obtained, an $(i,k)$-element of this matrix has value of 0 if $M^{(1)}(i,j)$ is equal to

|   | $a$ | $b$ | $c$ | $d$ | $e$ | $f$ | $g$ |
|---|---|---|---|---|---|---|---|
| $a$ | 0 | 0 | 0 | $acfd$ | $acfe$ | $abdf$ $acef$ | $abdg$ |
| $b$ | 0 | 0 | $bdfc$ | 0 | $bdfe$ $bace$ | $bacf$ | 0 |
| $c$ | 0 | $cfdb$ | 0 | $cabd$ $cefd$ | 0 | 0 | $cfdg$ |
| $d$ | $dfca$ | 0 | $dbac$ $dfec$ | 0 | $dfce$ | 0 | 0 |
| $e$ | $efca$ | $ecab$ $efdb$ | 0 | 0 | $ecfd$ | 0 | $efdg$ |
| $f$ | $fdba$ $feca$ | $fcab$ | 0 | 0 | 0 | 0 | 0 |
| $g$ | $gdba$ | 0 | $gdfc$ | 0 | $gdfe$ | 0 | 0 |

**Fig. 5.** $M^{(3)}$ matrix

|   | $a$ | $b$ | $c$ | $d$ | $e$ | $f$ | $g$ |
|---|---|---|---|---|---|---|---|
| $a$ | 0 | $acefdb$ | $abdfec$ | 0 | $abdfce$ | 0 | $acefdg$ |
| $b$ | $bdfeca$ | 0 | 0 | 0 | 0 | 0 | $bacfdg$ |
| $c$ | $cefdba$ | 0 | 0 | 0 | $cabdfe$ | 0 | 0 |
| $d$ | 0 | $dfecab$ | 0 | 0 | $dbacfe$ | $dbacef$ | 0 |
| $e$ | $ecfdba$ | 0 | $efdbac$ | $efcabd$ | 0 | $ecabdf$ | $ecabdg$ |
| $f$ | 0 | 0 | 0 | $fecabd$ | $fdbace$ | 0 | $fcabdg$ |
| $g$ | $gdfeca$ | $gdfcab$ | 0 | 0 | $gdbace$ | $gdbacf$ | 0 |

**Fig. 6.** $M^{(5)}$ matrix

0 or $\tilde{M}^{(1)}(j,k)$ is equal to 0 for every $j$. Zero occurs also in case where the $k$-node of the $\tilde{M}^{(1)}(j,k)$ occurs in the set of nodes of the element $M^{(1)}(i,j)$. When $M^{(1)}(i,j)$ element contains more than one sequence of nodes, then we proceed analogously obtaining more paths. Here in Figs. 4, 5 and 6 matrix containing all the paths of length equal to two, three and five (determined by the number of arcs) are presented, respectively.

**Table 2.** Numbers of straight paths of length $k$ from node $d$ to other destination nodes

| Destination node | | | | | | | | Sum of paths |
|---|---|---|---|---|---|---|---|---|
| $k$ | $a$ | $b$ | $c$ | $d$ | $e$ | $f$ | $g$ | |
| 1 | 0 | 1 | 0 | 0 | 0 | 1 | 1 | 3 |
| 2 | 1 | 0 | 1 | 0 | 1 | 0 | 0 | 3 |
| 3 | 1 | 0 | 2 | 0 | 1 | 0 | 0 | 4 |
| 4 | 1 | 1 | 0 | 0 | 1 | 1 | 0 | 4 |
| 5 | 0 | 1 | 0 | 0 | 1 | 1 | 0 | 3 |
| 6 | 0 | 0 | 0 | 0 | 0 | 0 | 0 | 0 |

Table 2 sets out the numbers of straight paths of length $k = 1, 2, ..., n-1$ from the source node to other nodes in the network, assuming that the node $d$ is the source node.

On the basis of the number of paths of a specified length $k$ from the node $d$ to the other nodes in the network it can easily determine the probability of getting information from the source node to the destination one. Given that each node in the network knows only its neighbours, sending packets along straight paths will require a modification of the H-P protocol consists in placing in the transmitted packet a list of all previously visited nodes. The last step is checking, in every relaying node, if the packet has not already been in the node that was choose to be the next in transmitting path.

### 3.3   The Second Modification of the H-P Protocol

Latin Multiplication can also be used to determine the directed routes $R = \{x_1, u_1, x_2, u_2, ..., u_{l-1}, x_l\}$ of any length - $l$, so that:

$$\bigwedge_{1<i<l} [\langle x_{i-1}, u_{i-1}, x_i \rangle \in R] \wedge [\langle x_i, u_j, x_{i-1} \rangle \notin R] \tag{1}$$

| | $a$ | $b$ | $c$ | $d$ | $e$ | $f$ | $g$ |
|---|---|---|---|---|---|---|---|
| $a$ | 0 | $acfdb$ | $abdfc$ $acfec$ $acefc$ | $acefd$ | $abdfe$ | 0 | $acfdg$ |
| $b$ | $bdfca$ | 0 | $bdfec$ | $bacfd$ | $bacfe$ $bdfce$ | $bacef$ | 0 |
| $c$ | $cfdba$ $cfeca$ $cefca$ | $cefdb$ | 0 | 0 | $cefce$ | $cfecf$ $cabdf$ | $cabdg$ $cefdg$ |
| $d$ | $dfeca$ | $dfcab$ | 0 | 0 | $dbace$ | $dbacf$ $dfecf$ $dfcef$ | 0 |
| $e$ | $efdba$ | $efcab$ $ecfdb$ | $ecfec$ | $ecabd$ | 0 | $efcef$ | $ecfdg$ |
| $f$ | 0 | $fecab$ | $fdbac$ | $fcabd$ $fecfd$ $fcefd$ | $fecfe$ | 0 | 0 |
| $g$ | $gdfca$ | 0 | $gdbac$ $gdfec$ | 0 | $gdfce$ | 0 | 0 |

**Fig. 7.** $M^{(4)}$ matrix

which means that the packet from the $x_i$ node may be directed to all adjacent nodes through the $u_j$ arc, except the node from which the packet was sent to the $x_i$ node. It should be noted that the Latin Multiplication allows to determine all the directed routes of a given length $l$ between all pairs of nodes in the graph. To make this possible, it should be assumed that the matrix element $M^{(l)}(i, k)$ will equal to 0 only when element $M^{(l-1)}(i, j)$ equals to zero or element $\tilde{M}^{(1)}(j, k)$

| | a | b | c | d | e | f | g |
|---|---|---|---|---|---|---|---|
| a | acefdba abdfeca | abdtcab acfecab acefcab | acfdbac | acfecfd | acfecfe | abdfecf abdfcef acefcef | 0 |
| b | bacfdba bacfeca bacefca | bdfecab bacefdb | bdfcefc | bdfcabd bdfecfd bdfcefd | bacefce bdfecfe | bacfecf | bacefdg |
| c | cabdfca | cfecfdb | cefdbac cfecfec cabdfec cefcefc | cfecabd cefcabd cefcefd | cfdbace cabdfce | cfdbacf | cfecfdg |
| d | dfcefca | dbacfdb dfecfdb | dbacfec dfecfec | dfecabd dbacefd | dfcefce | dfcabdf | dfcabdg dbacfdg dfecfdg |
| e | efcefca | ecfecab efcefdb | ecfdbac ecabdfc | ecfecfd | efdbace efcefce ecfecfe ecabdfe | efdbacf efcabdf | efcabdg efcefdg |
| f | fecfeca fecfdba fcefdba | 0 | ecfdbac ecabdfc | fdbacfd | fdbacfe fcabdfe | fecfecf fecabdf fdbacef | fecabdg |
| g | 0 | gdfecab | gdfcefc | gdfcabd gdbacfd gdfecfd gdfcefd | gdbacfe gdfecfe | gdbacef | 0 |

**Fig. 8.** $M^{(6)}$ matrix

equals to zero for every $j$ or both equal to zero. To realize the directed route of the length $l$, while still meeting the constraint (1), element $M^{(l)}(i,k)$ has to be equal to zero in the case when the last but one node, in the string of nodes of the element $M^{(l-1)}(i,j)$, is equal to $\tilde{M}^{(1)}(j,k)$ element. It should be emphasized that the introduction of the constraint (1) does not eliminate the possibility of a cycle in the directed route; cycles can occur, however, with a length greater than two. In Figs. 7 and 8 the directed routes having a length equal to 4, 6, respectively and meeting the constraint (1) are shown. It should be noted that the length of the two matrices $M^{(2)}$ in both cases is identical.

**Table 3.** Number of directed routes from node $d$ to destination nodes with the length of $k$, which meet

| Destination node | | | | | | | | Sum of paths |
|---|---|---|---|---|---|---|---|---|
| $k$ | a | b | c | d | e | f | g | |
| 1 | 0 | 1 | 0 | 0 | 0 | 1 | 1 | 3 |
| 2 | 1 | 0 | 1 | 0 | 1 | 0 | 0 | 3 |
| 3 | 1 | 0 | 2 | 0 | 1 | 0 | 0 | 4 |
| 4 | 1 | 1 | 0 | 0 | 1 | 3 | 0 | 6 |
| 5 | 0 | 1 | 1 | 3 | 2 | 1 | 0 | 8 |
| 6 | 1 | 2 | 2 | 2 | 1 | 1 | 3 | 12 |

Successive matrices, containing the directed route meeting the constraint (1) with a length of $k = 3, 5, 7, 8, 9, ...$ are not presented in this work. Table 3 shows the number of directed routes from node $d$ to other nodes in the network.

## 4    Obtained Results

Results shown in Tables 1, 2, 3 enable obtaining the probability of receiving information from a source node $d$ (in the general case, the source node can be adopted by any other node) to the destination nodes for assumed length of the route (paths in Table 2). This probability can be obtained by dividing the number of directed routes to the destination node by the total number of routes of predetermined length. In Fig. 9 the probability of achieving the node $e$ from the node $d$ having the length of the route/path 1-6 for each of the presented approach is presented. Based on this figure it can be said that the highest probability of reaching packet from node $d$ to node $e$ is when the packet travels along a straight path, but in this case the H-P protocol needs a bigger modification, which is based on the registration of all the nodes met on the route. The resulting probability values in this case can provide an upper estimation for the two other algorithms. In the case when the packet travels along the directed route meeting the constraint (1) the probability of receiving the packet by node $e$ is bigger for the most routes than for the routes without any limitations. In addition, the implementation of this case requires memorizing only the last but one node in the address field.

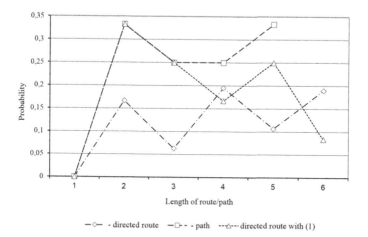

**Fig. 9.** The probability of achieving the node $e$ from the node $d$ having the length of the route/path 1, 2, 3, 4, 5, 6 for each of the presented approach

# 5  Summary and Conclusions

In this paper Latin Multiplication was used to analyze WSN network with Hot Potato protocol reading meters for the electric energy consumption. As a result, it is possible to determine the probability of the packet from acquisition node reaching a receiving node at a given number of steps analytically. In addition, there is a possibility of introducing additional constraints on routes/paths along which the sent packet moves. Using this method, due to the flexible ability of introducing constraints, has a definite advantage over the simulation methods, because it does not require multiple repetition of simulation runs for the estimation results. The obtained result for the seven-nodal network shows that the highest probability of reaching packet from source node to the destination node is when the package moves along the path. However, this requires the greatest interference in the Hot Potato protocol. The best results, from the practical point of view, requiring little interference in the Hot Potato protocol are obtained when the packet moves along the directed route with constraint that the node can send a packet to any successor, except the node which received the packet. Proposed in this work method may be implemented in acquisition nodes to decide which H-P variant is better for specific WSNs, taking into account their differences in topologies or in propagation conditions. Proposed method is not limited to WSNs, only. It can be implemented in any telemetry communication system that uses shared medium and needs multi-hop technique as a result of using short range devices as transceivers [7, 8].

# References

1. Al-Karaki, J.N., Kamal, A.E.: Routing techniques in wireless sensor networks: a survey. IEEE Wirel. Commun. **11**(6), 6–28 (2004)
2. Bieliński, W.: Struktura amplitudowo-fazowa profili obciążenia systemów elektroenergetycznych wybranych krajów. Rynek Energii (2)
3. Bononi, A., Tonguz, O.K., et al.: Analysis of hot-potato optical networks with wavelength conversion. J. Lightwave Technol. **17**(4), 525 (1999)
4. Dubalski, B., Kiedrowski, P.: WSN networks with hot potato protocol for automatic meter reading systems: Methods of analysis based on graph theory. Rynek Energii **5**, 48–53 (2010)
5. Goszczyński, T.: Określanie pokrycia bezprzewodowej sieci sensorowej metodą obliczania ścieżki najmniejszej ekspozycji. Pomiary Automatyka Robotyka **7**, 8 (2009)
6. Kaufmann, A.: Graphs dynamic programming and finite games. Technical report (1967)
7. Kiedrowski, P.: Toward more efficient and more secure last mile smart metering and smart lighting communication systems with the use of plc/rf hybrid technology. Int. J. Distrib. Sens. Netw. **2015**, 197 (2015)
8. Kiedrowski, P.: Errors nature of the narrowband PLC transmission in smart lighting LV network. Int. J. Distrib. Sens. Netw. **2016**, 3 (2016)

9. Kiedrowski, P., Dubalski, B., Marciniak, T., Riaz, T., Gutierrez, J.: Energy greedy protocol suite for smart grid communication systems based on short range devices. In: Choraś, R.S. (ed.) Image Processing and Communications Challenges 3. AISC, vol. 102, pp. 493–502. Springer, Heidelberg (2011)
10. Korzan, B.: Elementy teorii grafów i sieci: metody i zastosowania. Wydawnictwa Naukowo-Techniczne (1978)
11. Zhang, Z., Acampora, A.S.: Performance analysis of multihop lightwave networks with hot potato routing and distance-age-priorities. IEEE Trans. Commun. **42**(8), 2571–2581 (1994)

# Extreme Learning Machines for Web Layer Anomaly Detection

Rafał Kozik[1]([⊠]), Michał Choraś[1], Witold Hołubowicz[1,2], and Rafał Renk[2]

[1] Institute of Telecommunications,
UTP University of Technology and Science Bydgoszcz,
Bydgoszcz, Poland
rkozik@utp.edu.pl
[2] Adam Mickiewiecz University, UAM, Poznań, Poland

**Abstract.** The idea of service oriented architecture (SOA) and the wide adoption of the cloud computing cause the rapid advancement of web applications. Also the constantly increasing expectations of end-users concerning the usability of graphical interfaces have become a driving force for new information and communication technologies. However, as new technologies, frameworks and software solutions are created, it often happens that accidentally software flaws are introduced. In many cases, those flaws may have serious implications, such as privileges escalation, server and client sides infection with the malware or sensitive data leakage. Therefore, recent cyber incidents concerning web applications show that the new countermeasures are needed in order to protect the web layer. In this paper we propose the method that adapts the Extreme Learning Machine to solve the two class classification problem in the Web Layer Anomaly Detection domain. Our experiments give promising results proving that this technique can be used to effectively detect cyber attacks targeting web applications.

**Keywords:** Cyber security · Anomaly detection · Machine-learning

## 1 Introduction

Currently, it can be noticed that the increasing number of cyber criminals and hackers is getting interested in attacking and compromising the web-based applications. This comes from the fact that it is relatively easy to find a design or source code flaws in the web services. There are even search engines that help to find vulnerable web pages. There are different reasons why the attackers are interested in hacking an web page. Probably, what can be recently observed, the sensitive data (passwords, personal data, etc.) are of increasing value. Such data can be easily sold on black market, giving direct economic income and in result, the strong incentive and motivation for the criminals. Vulnerable pages can also be used to spread the malicious contents (viruses, worms, spam, etc.). Despite the fact that, there are tools [4,5,15], methodologies, and guidelines [1] helping the developers to address the security issues concerning web pages, still

R.S. Choraś (ed.), *Image Processing and Communications Challenges 8*,
Advances in Intelligent Systems and Computing 525, DOI 10.1007/978-3-319-47274-4_27

the vulnerable web applications are a common element in the vector of many cyber attacks [16,18].

Among all the attacks targeting web-servers, the SQLIA (SQL Injection Attack) still remains one of the most important network threat, which is ranked as one of the top threats on the OWASP list [1]. The code injection and other similar exploits are the results of interfacing a scripting language by directly passing information through another language and are ultimately caused by insufficient input validation. Particularly, the SQL Injection Attacks refer to a code-injection attacks category in which part of the user's input is interpreted as SQL code. A successful injection can cause serious consequences including data loss, corruption, lack of accountability or the denial of access. Additionally, the level of prevalence is described as common, while level of detectability is identified as average [11].

This paper is structured as follows. Firstly, we present the main building blocks of our method for detecting the cyber attacks. Afterwards, we present the detailed insight on key elements, namely data encoding and attacks detection by means of Extreme Learning Machines classifier. The evaluation methodology, experiments descriptions and results are given after. The paper is concluded with final remarks and plans for future work.

## 2    Related Work

There are several tools and methods that are dedicated to combat SQL injection attacks exploiting the above mentioned application flaws. Some of the frequently used tools use static code analysis approaches in order to find the vulnerabilities that may be exploited by any cyber attack. However, as it is stated in [11], the difficulty relates to fact that many kinds of security vulnerabilities are hard to be found automatically (e.g. access control issues, authentication problems). Therefore, currently such tools are able to automatically find a relatively small fraction of the application security flaws.

There are also solutions that adapt signature-based approach to describe (and detect) cyber attacks. Some examples include SCALP [4], PHPIDS [13], Snort [14]. The advantage of such tools is their ability to process huge amounts of data. This is due to the fact, that there are efficient algorithms that are able to check given piece of text against pattern (usually expressed as PCRE [12] regular expressions) in a short time.

However, the common drawback is that an expert knowledge is required to build such patters describing cyber attack. Moreover, such attacks like SQL injection are easy to obfuscate (e.g. using URL encoding). This makes the problem of providing reliable pattern of an attack difficult.

## 3    Proposed Method Overview

The architectural blueprint of the proposed method is shown in Fig. 1. The input for the attack detection procedure is the HTTP request sent from client to server.

Basing on such requests we want to model the content of the HTTP connection. In this approach, we make the assumption that the consecutive requests are statistically independent from each other. Hence, some attacks may exhibit patterns that sequentially appear over the time (e.g. scanning, probing, parameter tampering, sessions takeovers), and thus could be detected by correlating events in the time domain. However, we left that aspect for the future work. Therefore, we inspect each request and classify it as normal or anomalous when it arrives.

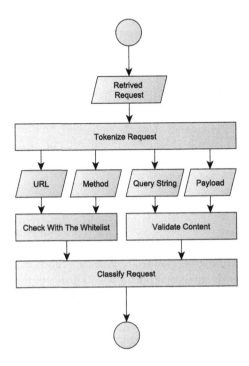

**Fig. 1.** The proposed algorithm overview. It adapts machine learning approach. First, the model is established on the learning data, afterwards new data samples are compared to the model ("Validate Content" block) in order to detect a cyber attack. The method returns binary result (A/N), where A indicates *anomaly* and N *normal*.

As it is suggested in [9], it is convenient to break each request into a sequence of so called tokens and values pairs. This concept comes from RFC2616 document, which describes HTTP protocol. An example of such a pair would be "*method = GET*" or "*content_type = html*". However, before we start the analyses of the request content, we apply simple categorisation. We group the requests by URL addresses and HTTP methods (e.g. PUT, POST, GET). This approach allows us to detect suspicious calls requesting unknown (not existing) resources. After the categorisation process we analyse the content of the request. The most probable areas of the request that may contain injected malicious code is either so called query string (commonly associated with the GET method), or request content. Therefore, we introduce here our machine-learning based approach for such validation.

# 4 Request Content Validation Algorithm

## 4.1 Extreme Learning Machine

The ELM (Extreme Learning Machine) is the method for the data classification that has been proposed by Huang et al. [7]. This approach extends the idea of single-hidden-layer neural networks (SLFNs) that are learnt without iterative process. The neurons in the hidden layer may not be necessarily the neurons in a classical sense, and they are not tuned during the learning phase. More precisely a response $f(x)$ for a signal $x$ is calculated using the following equation

$$f(x) = h(x)\beta \tag{1}$$

where $\beta$ indicates the output layer weights and $h(x)$ is the response of hidden layer to the input signal $x$. The $h$ layer performs random and non-linear projection. More precisely, each neuron in a hidden layer is fully connected to $d$ input signals through the set of randomly initialised weights $w$. Each neuron has also a bias $b$ which is also randomly selected. Therefore, the Eq. 1, for $L$ neurons in the hidden layer $h$, and some activation function $G$, will represented as

$$f(x) = \sum_{i=1}^{L} G(w_i, b_i, x)\beta_i \tag{2}$$

Therefore, for the whole training dataset $X, T = \{(x_j, t_j)\}_{j=1}^{N}$ (where $x_j \in R^d$ and $t_j \in -1, 1$), one can express the activation matrix as

$$H_{ij} = G(w_i, b_i, x_j) \tag{3}$$

From this point, we can formulate the Extreme Learning Machine optimisation problem as

$$\underset{\beta}{\text{maximise}} \ \|H\beta - T\|^2$$
$$\text{where} \quad H_{ij} = G(w_i, b_i, x_j), i = 1, ..., h, j = 1, ..., N \tag{4}$$

This can be solved by adapting state of the art linear algebra tools. As it is described in [8], the optimal solution in terms of mean squared error can be achieved using the Moore-Penrose (MP) pseudo-inverse (indicated as $H^\dagger$) of matrix $H$, such that $\beta$ can be calculated using formula

$$\beta = H^\dagger T \tag{5}$$

The $H^\dagger$ Moore-Penrose inverse matrix can be achieved using traditional orthogonal projection method, so that it can be calculated as

$$H^\dagger = (H^T H)^{-1} H^T \tag{6}$$

whenever $H^T H$ is non-singular or as

$$H^\dagger = H^T(H^T H)^{-1} \tag{7}$$

whenever $H H^T$ is non-singular. However, to achieve better stability and generalization performance, it is suggested to add positive value to the diagonal of $H^T H$ or $H H^T$ when calculating $\beta$, so that the equation will have the following form:

$$\beta = H^T(\frac{1}{\gamma} + H H^T)^{-1} T \tag{8}$$

Finally, the classification of the new data sample $x$ can be done using the following equation

$$C(x) = sign(\sum_{i=1}^{L} G(w_i, b_i, x)\beta_i). \tag{9}$$

## 4.2   Data Encoding

Hereby, as in our previous works [3,10], we have used typical approach for textual data encoding that adapts character distribution histograms. However, instead of classical one byte per bin association, we count the number of characters such that the decimal value in an ASCII table falls into particular ranges. Selected ranges represent different type of symbols like numbers, quotes, letters or special characters. We have distinguished the following ranges of characters:

- control characters (e.g. caret return, line feed, etc.) falling in range $[0, 31]$,
- SPACE!"#$%&'()*+,-./ characters falling in range $[32, 47]$,
- numbers contained in $[48, 57]$,
- :;<=>?@ characters falling in range $[58, 64]$,
- capital letters (range $[65, 90]$),
- [\]^_` characters falling in range $[91, 96]$,
- small letters (range $[97, 122]$)
- {~}| characters falling in range $[123, 127]$,
- other special characters falling in range $[128, 255]$.

In result our histogram will have 9 bins. This is significant dimensionality reduction in contrast to sparse 256-bin histograms (one bin per byte character). Moreover, we apply per token normalisation. More precisely, for each token, which is extracted from the query string or payload (see Fig. 1), we use the following formula:

$$V_i = \frac{(V_i - \overline{V_i})^2}{\overline{V_i}} \tag{10}$$

where $V_i$ indicates the $i$-th component of feature vector $V$, and $\overline{V_i}$ mean value of $V_i$ calculated for a given token, respectively.

# 5   Experiments

In this paper we use known benchmark dataset for web application cyber attacks and anomalies detection CSIC'10 [2] (quite recent one in comparison to other known datasets). However, it is already 5 years old and does not contain some of the recent web application anomalies. Therefore, we decided to enhance it with additional data coming from the monitoring of requests to real web application [10].

We intended to incorporate into the original dataset the traffic volume that will contain wider variety of HTTP requests, which were generated by more up-to-date web applications. The original dataset currently contains mainly requests that are complied with "application/x-www-form-urlencoded" encoding standard. However, the modern web applications are currently using also different types of request content, e.g. "applicaiton/json", "text/x-gwt-rpc", or any other proprietary types.

We increased the original CSIC'10 dataset by 3.6%. We added 2.08% of normal samples (around 1500 requests) and 7.9% of new attacks (around 2000 of anomalous requests). To gather these samples we set up a GWT-based web application that provides feature-rich GIS (Geographic Information System). In this application we identified several malicious code injection vulnerabilities. We used these vulnerabilities to convey cyber attacks.

# 6   Results

We have compared our method to other techniques that can be classified (according to provided in the introduction classification) as signature-based or anomaly-based. In all the cases we have measured the effectiveness of the methods using True Positives (TP) and False Positives (FP) ratios.

It can be noticed that signature-based methods achieve quite low detection ratios, while having low number of false positives. In our experiments, we have adapted two popular and often used application firewalls, namely PHPIDS and ApacheMod.

Moreover, we have compared our method to the compression-based method proposed in [17] and the approach adapting Nearest Neighbour search in Levenshtein distance [6]. In all cases our method can achieve better results, allowing us

**Table 1.** Comparison of effectiveness.

| Type | Method | TP rate | FP rate |
|---|---|---|---|
| Signature-based | PHPIDS | 0.2040 | 0.0125 |
| | ApacheMod | 0.2630 | 0.0034 |
| Anomaly-based | Compression | 0.4300 | 0.0000 |
| | LVSD | 0.6230 | 0.0010 |
| | Proposed method | 0.9498 | 0.0079 |

to detect almost 95 % of attack attempts while having low number of false positives (less than 1 % see Table 1). However, in case of ApacheMod, Compression-based technique, and LVSD the FP Ratios are lower, but the result for our method is comparable and very acceptable.

# 7  Conclusions

In this approach we propose the method that adapts the Extreme Learning Machine to solve the two class classification problem in the Web Layer Anomaly Detection domain. Our experiments have been conducted on the extended CSIC dataset that incorporates samples gathered for GIS web-based application that implements recent development concepts such as SPA (Single Page Application) and RIA (Rich Internet Applications) to provide cross platform and responsive user interface. Our experiments give promising results proving that this technique can be used to effectively detect cyber attacks targeting web applications while still having low number of false positives.

# References

1. Owasp top 10 2013 (2016). https://code.google.com/p/apache-scalp/
2. Giménez, C.T., Villegas, A.P., Marañón, G.Á.: The http dataset CSIC 2010 (2010). http://users.aber.ac.uk/pds7/csic_dataset/csic2010http.html
3. Choraś, M., Kozik, R.: Evaluation of various techniques for SQL injection attack detection. In: Burduk, R., Jackowski, K., Kurzynski, M., Wozniak, M., Zolnierek, A. (eds.) Proceedings of the 8th International Conference on Computer Recognition Systems CORES 2013, vol. 226, pp. 753–762. Springer, Switzerland (2013)
4. Apache Community: Apache log analyzer for security (2016). https://code.google.com/p/apache-scalp/
5. Damele, B., Stampar, M.: SQLMap: automatic SQL injection and database takeover tool (2015)
6. Gaikwad, S., Bogiri, N.: Levenshtein distance algorithm for efficient and effective XML duplicate detection. In: 2015 International Conference on Computer, Communication and Control (IC4), pp. 1–5. IEEE (2015)
7. Huang, G.B., Zhu, Q.Y., Siew, C.K.: Extreme learning machine: a new learning scheme of feedforward neural networks. In: Proceedings of 2004 IEEE International Joint Conference on Neural Networks, vol. 2, pp. 985–990. IEEE (2004)
8. Huang, G.B., Zhu, Q.Y., Siew, C.K.: Extreme learning machine: theory and applications. Neurocomputing **70**(1), 489–501 (2006)
9. Ingham, K.L., Inoue, H.: Comparing anomaly detection techniques for HTTP. In: Kruegel, C., Lippmann, R., Clark, A. (eds.) RAID 2007. LNCS, vol. 4637, pp. 42–62. Springer, Heidelberg (2007). doi:10.1007/978-3-540-74320-0_3
10. Kozik, R., Choraś, M.: Solution to data imbalance problem in application layer anomaly detection systems. In: Martínez-Álvarez, F., Troncoso, A., Quintián, H., Corchado, E. (eds.) HAIS 2016. LNCS (LNAI), vol. 9648, pp. 441–450. Springer, Heidelberg (2016). doi:10.1007/978-3-319-32034-2_37
11. OWASP: project homepage (2016). https://www.owasp.org/index.php
12. PCRE: perl compatible regular expressions (2016). http://www.pcre.org/

13. PHPIDS: project homepage (2016). https://github.com/PHPIDS/PHPIDS
14. SNORT: project homepage (2016). http://www.snort.org/
15. Owasp Team: Owasp zed attack proxy project (2016). https://www.owasp.org/
    index.php
16. TripWire: the talktalk breach (2016). http://www.tripwire.com/state-of-security/
    security-data-protection/cyber-security/the-talktalk-breach-timeline-of-a-hack/
17. Wang, N., Han, J., Fang, J.: An anomaly detection algorithm based on lossless com-
    pression. In: 2012 IEEE 7th International Conference on Networking, Architecture
    and Storage (NAS), pp. 31–38. IEEE (2012)
18. PC World: Havex malware variants target industrial control system and scada users
    (2016).   http://www.pcworld.com/article/2367240/new-havex-malware-variants-
    target-industrial-control-system-and-scada-users.html

# NEW QoS CONCEPT for Protecting Network Resources

Lukasz Apiecionek[✉], Jacek M. Czerniak, and Dawid Ewald

Department of Computer Science, Institute of Technology,
Kazimierz Wielki University, ul. Chodkiewicza 30, 85 064 Bydgoszcz, Poland
lukasz.apiecionek@ukw.edu.pl

**Abstract.** Distributed Denial of Service attacks are one of the main problem of computer networks. There is no any method for protecting network user from source of the attack. Such attack could block network resources for many hours, while existing methods for protecting networks are using only firewalls and IDS/IPS mechanisms. Such solutions are not enough nowadays. This article presents the concept of Quality of Services methods and some well know network protocols for preparing network to fight with the DDoS attacks. This proposed concept lets the administrator to protect their network resources during the attack.

## 1 Introduction

Everybody knows that IT systems are nowadays omnipresent and users need a fast access to information from every part of the network. Nowadays Distributed Denial of Service attacks have become a problem as they cause network unavailability by blocking services via seizing system resources in computers in the network until they stop working. A user who has already started working in the system loses the connection and cannot even log out of the system, which has to do it for him after the connection timeout is reached or when a broken connection is detected. DDoS attacks are nowadays a serious obstacle for IT systems efficient functioning and they have to be eliminated. Common methods of fighting the DDoS attack problems [2–4,9,12] are usually limited to using the Intrusion Detection System and Intrusion Prevention System (IDS/IPS in short) solutions. Such systems are efficient provided that they have a description of well know attacks or some kind of Artificial Intelligence solution which could learn the actions in some specific scenarios of attack. Other solutions suggest using a firewall mounted on the network edge. However, this firewall will only block the incoming traffic on specific ports or IP address ranges, which is not sufficient. This paper presents a concept of the Quality of Services mechanisms which implemented in routers could eliminate the DDoS attacks.

The structure of this paper is as follows. Section 2 shortly describes the issue of the DDoS attacks and introduces the proposed method for fighting them. Section 3 provides a conclusion and discussion over the developed method.

© Springer International Publishing AG 2017
R.S. Choraś (ed.), *Image Processing and Communications Challenges 8*,
Advances in Intelligent Systems and Computing 525, DOI 10.1007/978-3-319-47274-4_28

# 2   CONCEPT of QoS Method

## 2.1   Description of the DDoS Attacks

The DDoS attacks are widely described in the literature [4,5]. These attacks can be performed on various system resources: TCP/IP sockets [5,13] or DNS servers. Regardless of the method, the main principle is to simulate so many correct user connections that their number exceeds the actual system performance and drives it to abnormal operation. Papers [4,5,7–9] describe methods for dealing with the DDoS attacks by their global detection and the necessity of cooperation between network providers. The transmission of the attackers packets is done through the provider's network and if it cannot be blocked, it leads to data link saturation. Such saturation results in lack of connection to the server. The proposed solutions to prevent such situations are not specific and their implementation is associated with many problems. The most common concern is the limited performance of network devices. However, it is possible to limit the incoming traffic on a firewall and allow the servers to deal with the already established connection. This will let the users finish their work and the new users will be able to connect to the server. The QoS method implemented on routers are counting incoming traffic and decide which packet will be transferred to other network as first, and which will be the last. Such method are well known and implemented by network providers on their routers.

## 2.2   QoS Method Used on Routers

The QoS method implemented on routers are counting incoming traffic and decide which packet will be transferred to other network as first, and which will be the last. Such method are well known and implemented by network providers on their routers.

There were also some new QoS method ideas which could work on one routers and try to protect network resources locally [4]. But this solution will do not recognize the source of the attack and do not solve the problem. The hacker could still send their packet to the server.

Routers are exchanging lot of information between each other about reachability of the IP networks. This is done by routing protocols like OSPF, BGP or multicast routing protocols [10,11]. This mechanism could be used in new QoS method.

## 2.3   Proposition of the New QoS Method

Many QoS method are counting packets, but they do not know if the packet is a part of DDoS attack on some server. To fight with DDoS attack a new services for the network is required. Such services could use some well know mechanism like exchanging information between routers, SNMP protocols for getting another knowledge of traffic statistics. The proposition of the authors for new QoS Services which could works on routers has got a following steps:

- routers are collecting statistic of transferred traffic (1),
- statistics are divided into the counters of traffic to specific destination (2),
- routers are exchanging their statistic over SNMP (3),
- server which is an aim of the DDoS attack send a SNMP message to their router that it is under attack (4),
- routers are passing information between each other about the IP address of the aim of the attack (5),
- according to routers statistic they are looking for the source of the attack (6),
- when the sources of the attack are recognized, they are blocked (7).

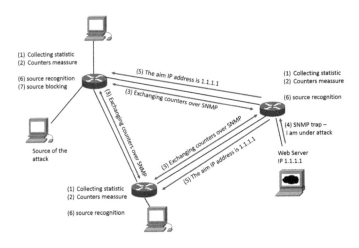

**Fig. 1.** QoS method concept in practice

This steps are presented on Fig. 1. This idea is very simple and could be implemented in easy way. The only thing to do is to implement it by network providers on their routers. All other proposed mechanism are well known.

### 2.4    Web Browsers Test

This method will not solve whole DDoS attack problems, but it will enable users to close their active connection when attack will start. Some test using Web browsers were made. Over 90 % of users used Internet Explorers (12 %), Mozilla Firefox (27 %) and Google Chrome (55 %) [1]. Using three most popular web browsers some test were made. Test procedure were to connect to web server which not exist and check how browser will send packets. Test condition:

- operating system Windows Vista,
- Internet Explorer version 9.0.8112.16421,
- Mozilla Firefox version 16.0.2,
- Google Chrome version 23.0.1271.64 m.

**Table 1.** Wireshark Web browsers connection test result

| Number of packets | Delays before next packet is send from browsers [seconds] | | |
| --- | --- | --- | --- |
| | Mozilla Firefox | Internet Explorer | Google Chrome |
| 1 | 0 | 0 | 0 |
| 2 | 0,254 | 0,001 | 0,001 |
| 3 | 2,996 | 2,995 | 0,25 |
| 4 | 3,246 | 2,995 | 2,996 |
| 5 | 8,997 | 8,995 | 2,996 |
| 6 | 9,247 | 8,995 | 3,246 |
| 7 | 20,996 | 20,995 | 8,997 |
| 8 | 21,239 | 23,991 | 8,997 |
| 9 | 21,246 | 29,992 | 9,248 |
| 10 | 23,995 | | |
| 11 | 24,235 | | |
| 12 | 24,245 | | |
| 13 | 29,998 | | |
| 14 | 30,238 | | |
| 15 | 30,248 | | |
| 16 | 42,231 | | |
| 17 | 45,23 | | |
| 18 | 51,233 | | |

Results of debug packets from Wireshark program are presented in Table 1. As it could be noticed, Mozilla Firefox browser tries to make a connection 18 times, while Google Chrome and Internet Explorer stops after 9 connection attempt.

Closing active connection and finishing work by users will be possible because of presented web pages browsers way of working. During browsers tests, authors recognize fact, that web page browsers made their work for user with some retransmission and they try to connect to web server more the once. Depending on browser which is chosen, there is nine chance to transfer appropriate data.

## 3   Conclusions

In this article a new concept of eliminating DDoS attacks was introduced. The methods suggested in the literature can block the access to the resources when the attack occurs, by using a firewall along with IDS/IPS mechanisms. During the time of the blockage no user from an external network can connect to the desired resources. The users who worked with the server lose their connection.

The method described in this article allows the network to find the sources of the attack. Then, such sources could be blocked and other users could still work with the server which was the aim of the attack. The part of this idea which has to be improved is a step 6 when routers are looking of the source of the attack. This is possible to do using for example fuzzy logic, which is described in literature [7]. Some other possibilities could be Kosinski's Fuzzy Numbers which are often better for some arithmetic reasons [6]. Besides the fact of the chosen algorithm for step 6, authors has a lot of idea which could be used [5,8]. This concept should be easy in implementation and during future test some algorithm will be implemented. Such idea should be considered to become some RFC standard. If there will be not any work on this topic, the DDoS attack will be huge problem in the future network.

# References

1. http://www.w3schools.com/browsers/browsers_stats
2. Cert advisory ca-1996-21 tcp syn flooding and ip spoofing attacks, November 2000. http://www.cert.org/advisories/CA-1996-21.html
3. Cert advisory ca-1996-01 udp port denial-of-service attack, September 1997. http://www.cert.org/advisories/CA-1996-01.html
4. Apiecionek, Ł., Czerniak, J.M., Zarzycki, H.: Protection tool for distributed denial of services attack. In: Kozielski, S., Mrozek, D., Kasprowski, P., Małysiak-Mrozek, B., Kostrzewa, D. (eds.) BDAS 2014. CCIS, vol. 424, pp. 405–414. Springer, Heidelberg (2014). doi:10.1007/978-3-319-06932-6_39
5. Czerniak, J.M., Apiecionek, Ł., Zarzycki, H., Ewald, D.: Proposed CAEva simulation method for evacuation of people from a buildings on fire. In: Atanassov, K.T., et al. (eds.) Novel Developments in Uncertainty Representation and Processing. AISC, vol. 401, pp. 315–326. Springer, Heidelberg (2016). doi:10.1007/978-3-319-26211-6_27
6. Czerniak, J., Dobrosielski, W., Apiecionek, Ł., Ewald, D.: Representation of a trend in ofn during fuzzy observance of the water level from the crisis control center. In: Proceedings of the 2015 Federated Conference on Computer Science and Information Systems, Annals of Computer Science and Information Systems, pp. 443–447. IEEE (2015)
7. Dickerson, J.E., Juslin, J., Koukousoula, O., Dickerson, J.A.: Fuzzy intrusion detection. In: IFSA World Congress and 20th NAFIPS International Conference, vol. 3, pp. 1506–1510. IEEE (2001)
8. Kozik, R., Choraś, M., Renk, R., Hołubowicz, W.: Semi-unsupervised machine learning for anomaly detection in HTTP traffic. In: Burduk, R., Jackowski, K., Kurzyński, M., Woźniak, M., Żołnierek, A. (eds.) CORES 2015. AISC, vol. 403, pp. 767–775. Springer, Heidelberg (2016). doi:10.1007/978-3-319-26227-7_72
9. Moor, D., Shannon, C., Brown, D.J., Voelker, G.M., Savage, S.: Inferring internet denial-of-service activity. ACM Trans. Comput. Syst. (TOCS) **24**(2), 115–139 (2006)
10. Piechowiak, M., Zwierzykowski, P.: The evaluation of multicast routing algorithms with delay constraints in mesh networks. In: 8th IEEE, IET International Symposium on Communication Systems, Networks and Digital Signal Processing CSNSDP, Pozna'n, Poland (2012)

11. Piechowiak, M., Zwierzykowski, P.: The evaluation of unconstrained multicast routing algorithms in ad-hoc networks. In: Kwiecień, A., Gaj, P., Stera, P. (eds.) CN 2012. CCIS, vol. 291, pp. 344–351. Springer, Heidelberg (2012). doi:10.1007/978-3-642-31217-5_36

12. Rocky, K., Chang, C.: Defending against flooding-based distributed denial-of-service attacks: a tutorial. IEEE Commun. Mag. **40**, 42–51 (2002)

13. Schuba, C.L., Krsul, I., Huhn, M.G., Spafford, E.H., Sundaram, A.: Analysis of a denial of service attack on tcp. Computer Science Technical Reports. Paper 1327 (1996). http://docs.lib.purdue.edu/cstech/1327

# QoS Mechanism for Low Speed Radio Networks - Case Study

Robert Palka$^{(\boxtimes)}$, Wojciech Makowski, Marcin Wozniak, Piotr Brazkiewicz, Krzysztof Wosinski, Pawel Baturo, Michal Terlecki, and Tomasz Gromacki

Teldat Sp.z o.o.sp.k ul., Cicha 19-27, Bydgoszcz, Poland
rpalka@teldat.com.pl

**Abstract.** A quick access to information is very important nowadays. Many systems exchange their data via radio links. These radio links are characterized by low bitrates and high bit error rate. In such situation, standard data exchange mechanisms cannot be used. In order to ensure the high quality for data transmission, other mechanisms have to be implemented. Such mechanism should be flexible and should adapt to the prevailing conditions of transmission. They should also allow to achieve optimum usage of the transmission medium. One of such mechanisms is the Battlefield Replication Mechanism. It has been adapted to work on mentioned radio links. This article presents the mechanisms for ensuring the delivery of the data used in the BRM - implemented and tested in practice.

**Keywords:** QoS · Radio replication mechanism

## 1 Introduction

Data communications systems that transmit data through low bit rates links are a special case. Such links are mainly used by the public services: the police, military, medical and crisis staff [7]. In such systems, data rate transfer can be decreased even to 1200 bps. Sending IP packets which size is 1400 Bytes can last over 9 s and depends on the protocol which is used. The time required for different size of packets on mentioned link is presented in Table 1.

At these speeds, transmission packet queues may fill up relatively quickly, depending on the amount of data to be transmitted in a particular application. With an average packet size of 768 bytes, the system has 5.1 s to analyze the packets in the queue and decide which packet should be transmitted in the first place. This gives an opportunity to optimize the queues of data packets.

If the transmission is done by radio links, the radio waves reach all receivers which are in the range of the transmitter. Such transmission usually works in broadcast mode. This gives an opportunity for optimization. If there is a need to send the same data packet to all recipients, there is no need to transfer them separately and simply only one single packet could be transmitted to all of the radios. Moreover, when the transmission is made only to selected recipients, this

© Springer International Publishing AG 2017
R.S. Choraś (ed.), *Image Processing and Communications Challenges 8*,
Advances in Intelligent Systems and Computing 525, DOI 10.1007/978-3-319-47274-4_29

**Table 1.** Time required for packet transferred over a link with data rate 1200 bps

| IP Packet size [Bytes] | Transmission time [s] |
|---|---|
| 51 | 0,4 |
| 255 | 1,7 |
| 510 | 3,4 |
| 768 | 5,1 |
| 1024 | 6,8 |
| 1400 | 9 |

can be done by indicating the recipients via the message in the radio network, but the transmission also is working in broadcast mode.

Data exchange in radio networks with low data rates is often characterized by a very high bit error rate. This is a challenge for the systems. One of the existing solutions which solve this problem is the Battlefield Replication Mechanism - BRM. The following sections of this article contains a description of this protocol with a mechanism which guarantees data packets delivery to the right destination.

## 2  A System with the BRM - Case Study

### 2.1  BRM Description

Battlefield Replication Mechanism is a data transmission protocol developed by engineers from TELDAT company [8]. BRM is using UDP (User Datagram Protocol), which is presented in Fig. 1.

As it is known, UDP uses a set of IP protocols and is a connectionless protocol which is not giving any guarantee that data packets will be delivered to the destination. The benefits of this protocol are: simplicity, lack of additional tasks (i.e. tracing session, establishing the connections) and the baud rate.

BRM protocol eliminates the disadvantages of using UDP, adds many new opportunities which improve its performance and ensures high security (encryption) of the transmission. BRM enables the exchange of operational data between databases (both version of the C2IEDM and JC3IEDM model can be used). BRM from its design phase was adapted to make the best and most effective work on low pass and unstable radio links, taking into account the current security requirements in ICT systems [1,2,4,5]. This protocol sends the minimum information required, ensures efficient bandwidth usage, adapting transmission parameters to the constantly changing environment. In order to ensure a high level of security of the transmitted information during various missions, the entire transmission is encrypted, and the data are further grouped, filtered and compressed (these treatments can increase the efficiency of transmission).

Each data transmission is encrypted using a symmetric key generated for this transmission. This key is exchanged (using the method of secure key exchange)

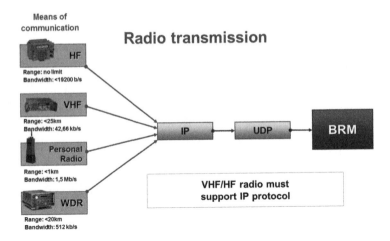

**Fig. 1.** Radio transmission with BRM protocol

between the points of replication during the connection phase, and is known only to the parties directly enumerating the data. Of course there is the ability to dynamically change it during runtime.

The most important features of the BRM protocol:

– uses a well-known and standardized UDP protocol;
– provides the delivery confirmation data;
– ensures high security - data encryption;
– operates on a low-throughput radio links;
– ensures an efficient use of the transmission links - additional compression;
– adatps the transmission depending on the transmission condition;
– provides a secure exchange of the encryption key;
– enables an automatic renewal of the encryption key;
– has a build-in mechanism for eliminating errors - integrity of the operational data;
– enables the replication of the data between C2IEDM and JC3IEDM databases of MIP program (sending the minimum required amount of data without loss of information).

## 2.2   Mechanisms Implemented in BRM

Sending information through the BRM mechanism consists of several steps, giving the assurance of delivering data (through the retry message). Data exchange mechanism guarantees delivering the data to the point of replication to which the connection exists. A confirmation mechanism is shown in Fig. 2. In this scenario, it is assumed that the first mobile vehicle is going one way, and at this time the BRM sends the position to the second mobile vehicle in the following manner:

- position is transmitted as a data packet called light, which means that it consists of minimal amount of data for passing information about the position and movement,
- if the packet will not reach the destination (second vehicle) in the configured amount of time (according to data errors or some other problems in the transmission) which in the presented situation lasts 15 s and is called AwaitingTimeOut, there will be a retransmission started till the situation when vehicle 1 will get an acknowledge from the second vehicle.

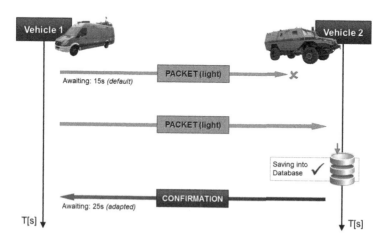

**Fig. 2.** Data confirmation mechanism

BRM is also equipped with a mechanism of missing data analysis and their replenishment. Figure 3 illustrates the operation of the mentioned mechanism. In the same situation as mentioned previously:

- the first vehicle sends a packet in light version(which consist of minimum information about position and movement) to the second vehicle over the radio link;
- second vehicle tries to save the data into the database;
- if the procedure of saving data fails (the second vehicle will not have the information about the first vehicle), there will be a special packet generated, which informs the first vehicle that the procedure of saving data failed;
- if the first vehicle gets the information about problems in saving data, it will generate new packets which consist of all of the required information for saving data into the database.

The mechanism of data exchange can operate with any radio station: from HF via VHF, Personal Radio to Wide Digital Radio. It should be noted that the radio used to transmit data, has to support IP and UDP transmission, which the BRM protocol uses.

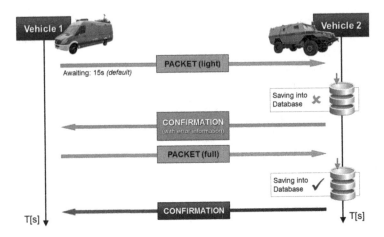

**Fig. 3.** Data analysis mechanism in BRM

The format of BRM package is shown in Fig. 4. In order to optimize the performance, the BRM protocol sends the data as tasks and identifies the appropriate task by the header added in the package.

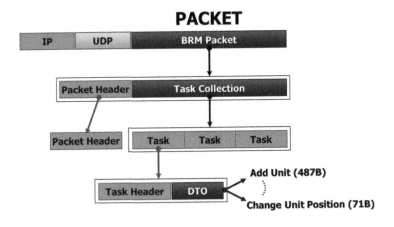

**Fig. 4.** BRM package structure

In accordance with the principle of this operation, the data can be transferred in full, standard or limited version. The types of packages are shown in Fig. 5.

If the average packet size is about 136 bytes and the transmission speed is at 1200 bps, the packet transmission time is 906 ms. The transmission system has got almost one second in average for the optimization of the queues.

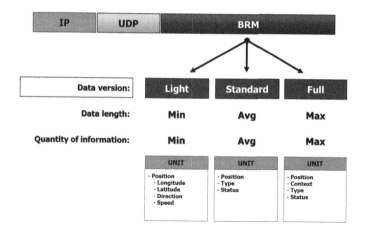

**Fig. 5.** Types of BRM packages

## 2.3   Queues Mechanism in BRM

The BRM protocol uses queue optimization mechanisms, based on the experience gained in the study of routing protocols [9]. The first step of the optimization is dropping packets consisting the same information. This could be made because packet flow to the queue could be faster than the transmission from the queue over the radio link. The transmission of the packet which consist the same data is unnecessary.

When the radio works in broadcast mode, all of the receivers get the same data. There is no requirement for sending packets in unicast mode if the packets are consisting the same data. In such situation the queue is optimized and data are exchanged by only one packet which consists the same information. This mechanism is changing unicast transmission to multicast mode and works also in some specific situations in broadcast mode.

The above mentioned mechanisms are used for queues optimization which lets us save data throughput and is separated from the application layer in that fact, so there are no requirements for any additional task to do in the application. The above mentioned mechanism could be also used in any type of transmission medium.

## 3   Conclusions

This article presents the practical implementation of the transmission mechanism of data within links with low data rates. This mechanism is using optimization procedures for the queues. The devices have got a lot of time for queues optimization in the network with low data rates links. When they poses some free computing power it gives very good results, while simple packet dropping can cause increasing of data flow and finally blocking the network. BRM mechanism is widely used in data transmission for the military systems [6] and also could be successfully used for communication with aircrafts [3].

# References

1. Apiecionek, Ł., Czerniak, J.M., Zarzycki, H.: Protection tool for distributed denial of services attack. In: Kozielski, S., Mrozek, D., Kasprowski, P., Małysiak-Mrozek, B., Kostrzewa, D. (eds.) BDAS 2014. CCIS, vol. 424, pp. 405–414. Springer, Heidelberg (2014). doi:10.1007/978-3-319-06932-6_39
2. Apiecionek, Ł., Makowski, W.: Firewall rule with token bucket as a DDoS protection tool. In: 2015 IEEE 13th International Scientific Conference on Informatics, pp. 32–35. IEEE (2015)
3. Apiecionek, Ł., Makowski, W., Biernat, D., Łukasik, M.: Practical implementation of AI for military airplane battlefield support system. In: 2015 8th International Conference on Human System Interaction (HSI), pp. 249–253. IEEE (2015)
4. Apiecionek, Ł., Motylewski, R., Stosik, P.: Bezpieczenstwo transmisji danych w systemach monitorowania wyposażenia straży pożarnej, problemy monitoringu eksploatacji sprzetu i wyposażenia straży pożarnej. CNBOP-PIB, pp. 41–48 (Jozefow 2015). doi:10.17381/2015.2
5. Apiecionek, L., Romantowski, M.: Security solution for cloud computing (2014)
6. Kruszyński, H.: Możliwości mobilne systemu JASMIN. Wsparcie teleinformatyczne dowodztw w dzialaniach wojsk ladowych, pp. 119–129
7. Kruszyński, H., Kosowski, T.Z.: Zarzadzanie wsparciem wspolpracy cywilno-wojskowej w dzialaniach kryzysowych, public managmenet 2013, wyzwania i dylematy zarzadzania organizacjami publicznymi 1, pp. 521–542 (2013)
8. Muchewicz, K., Sierakowski, L.: Sposoby wymiany danych operacyjnych w systemie jasmin, materiały konferencyjne xvii automatyzacja dowodzenia (2009)
9. Piechowiak, M., Zwierzykowski, P., Hanczewski, S.: Performance analysis of multicast heuristic algorithms. In: Third International Working Conference on Performance Modelling and Evaluation of Heterogeneous Networks, p. 41. Networks UK Publishers (2005)

# IoT WiFi Home Network Stress Test

Piotr Lech$^{(\boxtimes)}$ and Przemysław Włodarski

Department of Signal Processing and Multimedia Engineering, West Pomeranian University of Technology, Sikorskiego 37, 70–313 Szczecin, Poland
{piotr.lech,przemyslaw.wlodarski}@zut.edu.pl

**Abstract.** The article presents the research results of home WiFi network, designed to connect devices of the Internet of Things. Conducted stress tests had the objective to determine the parameters of the network and to what extent can it be integrated with other services. A particularly important aspect of a converged network is to ensure the comfort of people who access to web services and reliable communication in the machine-to-machine relationship. The study should help to determine whether sharing of a single network between IoT devices and the typical family access to Internet services is possible.

**Keywords:** Network stress test · IoT · WiFi

## 1 Introduction

The construction of convergent networks, providing the ability to share different services, realized on the basis of IP technology is the new trend [1] of development illustrated in Fig. 1. Observed growth of interest in WiFi technology in the IoT applications caused greater access of cheap communication modules with low power consumption [4]. Modules such as System-on-a-Chip have simplified the process of designing wireless IoT devices in relation to modular devices in which part of the connection is separated from the microprocessor. Formed IoT devices-network traffic is insignificant in relation to other sources of data transmission in IP network as shown in Table 1.

However, the location of the installation of IoT devices in one common space with other services can result in adverse interactions, such as loss of measurement data, delays in control etc. An important element in the design of reliable communication infrastructure in the converged sensor networks input is to know its potential and behavior in critical situations (congestion, failures). One of the methods for determining these parameters is a stress test (Table 2).

Observation of the network status is the basis for determining its quality using measurements associated with the following data transfer parameters in the network such as Latency or Throughput Fault evidence.

Latency reflects the speed of the network. The lower the latency is the faster the network works. The packet of data travel between the sensor node and the gateway node through the network. The amount of data that passed through

© Springer International Publishing AG 2017
R.S. Choraś (ed.), *Image Processing and Communications Challenges 8*,
Advances in Intelligent Systems and Computing 525, DOI 10.1007/978-3-319-47274-4_30

**Table 1.** Comparison of throughput of Internet applications

| Application | Throughput |
|---|---|
| IoT low power | 100 bps–100 Kbps |
| Web - access | 60–500 Kbps |
| Audio | 100–1000 Kbps |
| Video | 1–6 Mbps |
| File Sharing | 2–8 Mbps |

**Table 2.** Quality requirements for various classes of network services

| Class network services | Delay | Jitter | Services |
|---|---|---|---|
| Real Time (RT) | 100 ms | 50 ms | Video, IoT RT devices |
| Time Critical | 400 ms | unimportant | data control, IoT control |
| Non Real Time | 1 s | unimportant | Web services, IoT Beacon |
| Best Effort | unimportant | unimportant | e-mail |

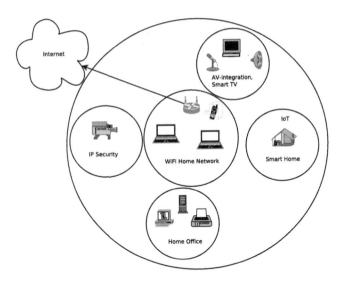

**Fig. 1.** New home WiFi concept

network (in segment time e.g. per seconds) is a useful information about potential of stable data transfer. The high throughput is required for video streaming with quality of service. Another important information about network is the degree of fault tolerance or fault resilience. Fault tolerance gives us information on the stable operation of any server services which should not notice any failure. Fault resilience gives us information about a fault which may be observed only for uncommitted data.

## 2   WiFi Internet of Things Architecture

We can encounter various options for convergence networks and services, however, if viewed from the point of view of used technologies of access devices, they
can be divided into two categories of accessibility to services depending on the
location of the server:

– services are available locally
– services are available globally.

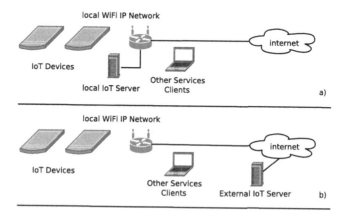

**Fig. 2.** WiFi IoT architecture: (a) LAN server localization, (b) WAN server localization

The first of them contains the categories in local network and the second one
is within the scope of the global network (Fig. 2). For these variants one should
adjust the test procedures.

## 3   Network Stress Test

There are several types of stress tests, for example tests related to the definition
of uptime or network devices associated with the reliability of data transmission.
The first one requires climatic chambers and flow generators of large amounts
of data. The second one opens network congestion similar to a DoS (Denial of
Service) attack [6], creating a data transfer to test the throughput of the devices
to connect or services. The tests associated with the generation of packet flow in
the network are used to determine the behavior of the network in the following
three categories: Latency, Throughput and Fault evidence [7].

We considered two sample topologies, one for the local network and the other
one involving a global network as shown in Fig. 2.

In the experiment 14 Intel Galileo 2 microcontrollers playing the role of the
IoT devices have been used, together with the Access Point with router - Banana

Pi R1, which plays the role of gateway providing access to the Internet. The IoT services server provides the IoT in the form of a micro-computer based on the Orange Pi. The topology of WAN connection is presented in Fig. 2, where server which provides the IoT is on the side of the global network. This structure reflects the latest trends in IoT applications, where IoT control services has been moved from local network to remote services in the cloud.

For the measurement and generation of network traffic [5] IPerf and JPerf tools have been used (available at https://iperf.fr). The JPerf works in graph mode and it is useful to visualize the measurement data. This program was selected among alternative programs, which characteristics can be found at https://www.caida.org/tools/taxonomy/perftaxonomy.xml.

The network link is delimited by the pair of hosts - client and server - with running IPerf software for client and JPerf for server. Client software has been installed on all IntelGalileo IoT hosts, server only on Orange Pi IoT Services Server. The server was connected using the wired network in the Fast Ethernet 100BaseTX standard. Access to a global network has been implemented in the standard Fast Ethernet 1000BaseT. The ISP restricts the bandwidth at 6 Mb/s inbound and 1 Mb/s for outgoing data.

All of IoT devices have been linked with AccessPoint within short 3 m distances (Fig. 3).

**Fig. 3.** All strong links

This software is installed as real time recorder with complete automatization of measurements and processing of stored data.

An additional advantage of the selected measurement tools is the ability to running in parallel multiple instances of a program that allows you to generate traffic that simulates using several sensors connected to a single IntelGalileo 2 at the same time. IPerf is ideal for performing stress tests because the tool is trying to use all of the available bandwidth to transfer data.

The state of network quality with connected link can be tested as follows:

- by Ping command for Latency with response time,
- IPerf UDP (typical protocol for sensors usage) test for Jitter with latency variation,
- IPerf UDP test for bandwidth throughput and datagram loss,

– IPerf TCP (typical protocol for actuator usage) test for bandwidth throughput and transmission errors.

Observation for each topology was held according to the following scheme shown in Fig. 4.

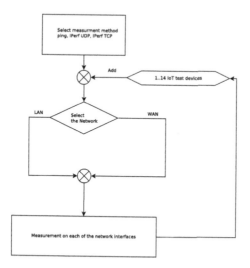

**Fig. 4.** Measurement test scheme

The results of data throughput tests (measurements on every IoT WiFi interface) for all IoT devices with IoT Server Services are shown in Table 3 for LAN and in Table 4 for WAN.

**Table 3.** The values of the measurements carried out in a local area network (link: server - IoT device, UDP - throughput, Ping - delay)

| Device No | 1 | 2 | 3 | 4 | 5 | 6 | 7 | 8 | 9 | 10 | 11 | 12 | 13 | 14 |
|---|---|---|---|---|---|---|---|---|---|---|---|---|---|---|
| Max Throughput [Kbps] | 456 | 506 | 459 | 393 | 395 | 499 | 579 | 498 | 599 | 410 | 391 | 442 | 432 | 514 |
| Average Thr. [Kbps] | 325 | 239 | 299 | 249 | 235 | 201 | 422 | 298 | 379 | 210 | 211 | 212 | 239 | 314 |
| Max Delay [ms] | 10 | 20 | 18 | 45 | 15 | 32 | 43 | 18 | 39 | 10 | 32 | 21 | 19 | 24 |
| Min Delay [ms] | 2 | 4 | 3 | 4 | 3 | 4 | 2 | 5 | 4 | 3 | 5 | 5 | 5 | 4 |
| Average Delay [ms] | 8 | 7 | 6 | 7 | 6 | 6 | 9 | 8 | 5 | 5 | 7 | 6 | 5 | 6 |

## 3.1   Throughput and Jitter Test

During the experiments three series of measurements have been made; each of them had to define the interactions between the IoT devices of the network and

**Table 4.** The values of measurements carried out in a wide area network (link: server - IoT device, UDP - throughput, Ping - delay)

| Device No | 1 | 2 | 3 | 4 | 5 | 6 | 7 | 8 | 9 | 10 | 11 | 12 | 13 | 14 |
|---|---|---|---|---|---|---|---|---|---|---|---|---|---|---|
| Max Throughput [Kbps] | 152 | 121 | 140 | 189 | 123 | 121 | 176 | 128 | 211 | 174 | 151 | 172 | 158 | 154 |
| Avarage Thr. [Kbps] | 120 | 111 | 123 | 129 | 115 | 101 | 122 | 118 | 109 | 130 | 114 | 112 | 139 | 124 |
| Max Delay [ms] | 508 | 567 | 499 | 565 | 536 | 432 | 632 | 475 | 432 | 534 | 442 | 456 | 669 | 544 |
| Min Delay [ms] | 41 | 53 | 123 | 222 | 33 | 24 | 22 | 15 | 24 | 13 | 35 | 25 | 19 | 14 |
| Avarage Delay [ms] | 80 | 74 | 69 | 78 | 65 | 63 | 99 | 58 | 75 | 55 | 87 | 62 | 59 | 55 |

availability of the services. Measurement data clearly show a downward trend in the transmission between the positions of the IoT and WiFi router, as a result of sharing of the limited wireless resource and some problems with efficient routing (Figs. 5 and 6).

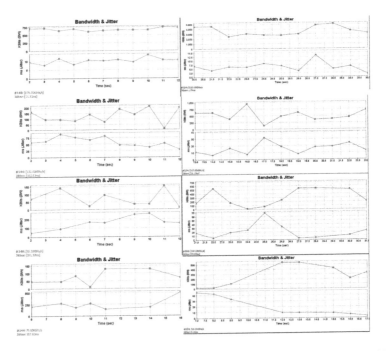

**Fig. 5.** Snapshot of JPerf UDP bandwidth and Jitter tests (left column - WAN connection, right column - LAN connection, 1st row - 1 IoT interface, 2nd row - 5 IoT interfaces, 3rd row - 10 IoT interfaces, 4th row - 14 IoT interfaces)

## 3.2 Latency Test

Examination of delays in the network was performed independently from the other tests, providing that the study conducted in parallel, using the Ping tool,

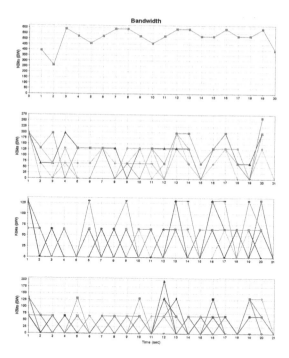

**Fig. 6.** Snap shot of JPerf TCP Fault Test (1st row - 1 IoT interfaces, 2nd row - 5 IoT interfaces, 3rd row - 10 IoT interfaces, 4th row - 14 IoT interfaces)

may introduce errors in the measurements. Delay test is particularly important in the case of extensive network delays because it has much greater value in relation to delays in local (Table 3) or wide networks (Table 4), because it allows you to determine the suitability of the compounds for the management of the IoT devices with the external server services.

### 3.3   Fault Test

Examination of frequency of occurrence of transmission errors involved the determination of the number of accidents and their duration. Considerable number of transmission errors has grown along with the number of connected IoT devices, the cause of this phenomenon is the router, which is unable to handle such a large network traffic (Table 5).

**Table 5.** Time failure shared in total transmission time for WAN and LAN - TCP connection registered on server interface.

| Connection | 1IoT | 5IoT | 10IoT | 14IoT |
|---|---|---|---|---|
| WAN | 3.3 % | 34.1 % | 63.2 % | 85.0 % |
| LAN | 0.0 % | 20.1 % | 34.1 % | 79.1 % |

# 4   Conclusions

The stress tests which had been carried out allowed to check available capacity and reliability of the network. At the same time they ensured registration of data about delays and jitter data transmission in the network. In the proposed example of network the weakest element turned out to be the Banana Pi R1 router, because it caused serious problems for a large amount of network traffic. It has been observed for the case of a smaller number of connected devices, that IoT network will be able to coexist with the home network, assuming that only a low amount of traffic will be generated by the IoT devices. The network testing has demonstrated a clear deterioration of stability with increasing number of the IoT devices.

# References

1. Al-Ali, A.R., Aburukba, R.: Role of internet of things in the smart grid technology. J. Comput. Commun. **3**(05), 229 (2015)
2. Amadeo, M., Campolo, C., Iera, A., Molinaro, A.: Information centric networking in IoT scenarios: The case of a smart home. In: 2015 IEEE International Conference on Communications (ICC), pp. 648–653. IEEE, June 2015
3. Bholebawa, I.Z., Jha, R.K., Dalal, U.D.: Performance analysis of proposed openflow-based network architecture using mininet. Wireless Pers. Commun. **86**(2), 943–958 (2016)
4. Choubey, P.K., Pateria, S., Saxena, A., Chirayil, S.B.V.P., Jha, K.K., Basaiah, P.M.S.: Power efficient, bandwidth optimized and fault tolerant sensor management for IOT in Smart Home. In: 2015 IEEE International Advance Computing Conference (IACC), pp. 366–370. IEEE, June 2015
5. de Veciana, G., Andrews, J.G., Shakkottai, S.: IEEE 802.11 wireless LAN traffic analysis: a cross-layer approach (2005)
6. Poddar, V., Jaipur, R., Chopra, M.A.: Detection of the de-authentication denial of service attack in 802.11 wireless networks. Int. J. Sci. Eng. Res. **6**(10), 150–158 (2015)
7. Zhuang, Y., Cappos, J., Rappaport, T.S., McGeer, R.: Future Internet Bandwidth Trends: An Investigation on Current and Future Disruptive Technologies. Technical Report TR-CSE-2013-0411/01/2013, Polytechnic Institute of NYU, Department of Computer Science and Engineering (2013). http://www.cs.ubc.ca/yyzh/tr-cse-2013-04.pdf. Accessed 22 Feb 2014

# Determining the Bit Error Rate
# for Redundant Transmission

Przemysław Włodarski$^{(\boxtimes)}$ and Piotr Lech

Faculty of Electrical Engineering, Department of Signal Processing and Multimedia
Engineering, West Pomeranian University of Technology, Sikorskiego 37,
70-313 Szczecin, Poland
{przemyslaw.wlodarski,piotr.lech}@zut.edu.pl

**Abstract.** In this article a bit error rate for redundant transmission is
determined. The method is based on the error rate of received bitstream
for different levels of redundancy and does not require any information
about transmission channel and original data. The main goal was to
derive a simple closed form expression that leads to good performance
and makes the calculation of bit error rate easy to implement in the low
power receiver nodes for wireless sensor network.

**Keywords:** Bit error rate · Retransmissions · Redundancy

## 1 Introduction and Motivation

The vast majority of current and future communication applications are exposed
to a number of factors such as: electromagnetic interference (EMI), thermal noise, signal distortion during transmission (attenuation), crosstalk, echo,
impulse noise (e.g. engines, relays), phase jitter etc. These factors can cause bit or
packet losses, especially across heavily loaded links. The most popular methods
to minimize losses are based on timeouts and retransmissions of messages that
are not acknowledged by the receiver. They are derived from Automatic Repeat
reQuest (ARQ) and are widely used i.e. in Ethernet (TCP protocol, ITU-T G.hn
specification) [4,9,10]. Unfortunately ARQ cannot be used in transmit-only systems consisting of many transmitters (sensors) and one receiver (gateway node),
like in topologies and systems appropriate for low cost wireless sensor networks
(WSN), the key technology for Internet of Things (IoT) [1,2,7].

Typically, a tiny-sized and limited-power sensor transforms light, vibration,
sound and chemical signals into digital data and then transfers them to the
microcontroller which processes the data and sends it through the wireless transmitter (RF module) to the the sink node. To decrease bit error rate (BER) in
this configuration, a Forward Error Correction (FEC) technique can be used [3].
FEC provides error correction without retransmissions and reduces the cost of
the sensor network elements. It can be classified as block codes and convolutional codes, where block codes are the most popular error correction codes, for

© Springer International Publishing AG 2017
R.S. Choraś (ed.), *Image Processing and Communications Challenges 8*,
Advances in Intelligent Systems and Computing 525, DOI 10.1007/978-3-319-47274-4_31

example: repetition, Hamming, Reed-Solomon, low density parity check (LDPC) or turbo product codes [5,6,8].

The simplest FEC block code method is the repetition $(n, 1)$ code, where each data bit is transmitted $n$ times. It is a relatively inefficient method from the point of view of bandwidth utilization, however, it has better reliability than Hamming codes if the bit error transmissions are bursty (e.g. fading channel) or if the signal-to-noise ratio (SNR) is very low. Therefore, in this article a method of determining of BER in this type of FEC for WSN, where each sensor has simple and cheap (less than \$1) transmitter and works in one sink node environment, is investigated.

## 2   Derivation and Results

BER is the relation of number of received error bits for a given number of transmitted bits and is denoted here by $p$, where $p \in [0, 1]$. It should have small values in general (typically less than $10^{-6}$), except for low SNR. There is a very simple relationship between bit error rate and packet error rate (PER) represented by the following formula:

$$P_{PER} = 1 - (1 - p)^N,  \tag{1}$$

where N is the packet size. This relationship can be approximated, but only for small bit error rates, by:

$$\tilde{P}_{PER} = p \cdot N.  \tag{2}$$

Looking from the point of view of reliability, Figs. 1 and 2 show the probability that a packet is received correctly, which corresponds to $1 - P_{PER}$. One can see that only for very small values of $p$, the relationship for both $P_{PER}$ vs $p$ and $P_{PER}$ vs $N$ is almost linear.

It is assumed that the original packet is not known and the bit errors do not affect the packet length. Redundant packets are received and then compared with each other, using XOR operation. If there is a difference in a given bit position, an error is registered. Intuitively, the probability of the registered error should be $n \cdot p$, where $n$ is the number of redundant packets, but it is not, because all bits in the same position could be changed. One can write an expression for the probability that minimum one bit is different for $n$ received packets, i.e. the probability that not all bits are changed or unchanged:

$$q = 1 - [(1 - p)^n + p^n]  \tag{3}$$

One can easily calculate a tangent for the above expression, determining a derivative for $p = 0$:

$$q' = p \cdot n \cdot \left[(1 - p)^{n-1} - p^{n-1}\right]\big|_{p=0} = p \cdot n,  \tag{4}$$

which is the same as the intuitive expression value mentioned earlier. Unfortunately, the relationship between BER and bit error detection is highly nonlinear

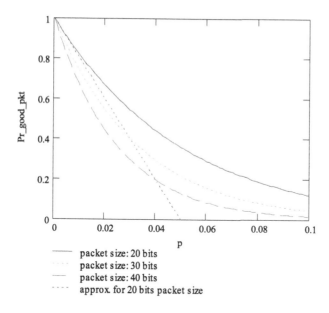

**Fig. 1.** Probability of correct packet reception versus BER for different packet sizes

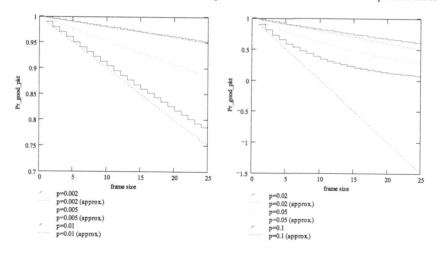

**Fig. 2.** Relationship between probability of correct packet reception and packet sizes for different BER values (smaller on the left, bigger on the right)

(Fig. 3) and, as can be clearly seen in Figs. 4 and 5, $q$ cannot be approximated by this linear relationship due to the big discrepancies, especially for high values of $p$ or $n$.

It is difficult to find a solution for $p$ directly from (3), but the $(1 - p)^n$ term can be extracted to binomial series $B(p, n)$ and the expression for $q$ can be rewrited as:

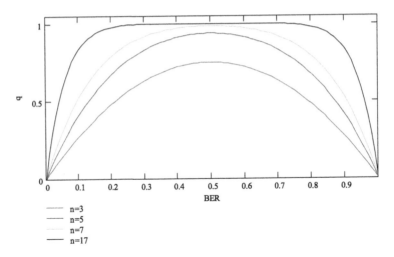

**Fig. 3.** Probability of incorrect bit detection against BER for different levels of redundancy

$$q = 1 - B(p, n) - p^n, \tag{5}$$

where $B(p, n)$ is the finite sum, since $n$ is an integer, and has the following form:

$$B(p, n) = \sum_{k=0}^{n} \binom{n}{k} \cdot (-1)^k \cdot p^k = 1 - n \cdot p + \frac{n \cdot (n-1)}{2} \cdot p^2$$

$$+ -\frac{n \cdot (n-1) \cdot (n-2)}{6} \cdot p^3 + \cdots + n \cdot (-1)^{n-1} \cdot p^{n-1} + (-1)^n \cdot p^n. \tag{6}$$

Substituting (6) into (5) one can get another expression (that gives almost exact values of $q$) for the base equation in (3), gaining four terms when summing up:

$$\tilde{q} = \sum_{k=1}^{n-1} \frac{n! \cdot (-1)^{k-1} \cdot p^k}{k! \cdot (n-k)!} = q - \tilde{e}, \tag{7}$$

where correction $\tilde{e}$ is only present for even values of $n$:

$$\tilde{e} = p^n \cdot \left[ (-1)^{n-1} - 1 \right]. \tag{8}$$

In order to determine bit error rate based on $q$ value, it is proposed to take two first terms of $\tilde{q}$:

$$\hat{q} = \frac{n \cdot \hat{p} \cdot (2 - n \cdot \hat{p} + \hat{p})}{2}. \tag{9}$$

Then, solving quadratic equation for the value of $\hat{p}$:

$$\hat{p}^2 \cdot [n \cdot (n-1)] + 2 \cdot n \cdot \hat{p} - 2 \cdot \hat{q} = 0, \tag{10}$$

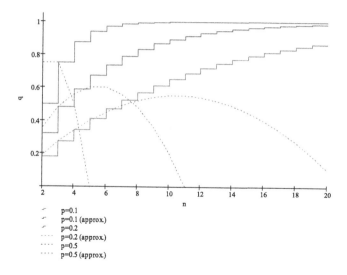

**Fig. 4.** Probability of incorrect bit detection for higher BER

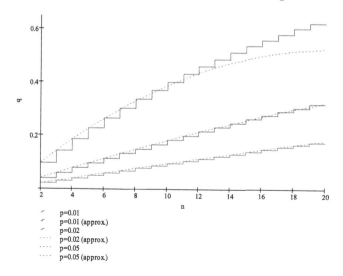

**Fig. 5.** Probability of incorrect bit detection for lower BER

one obtains the following solutions:

$$\hat{p} = \frac{1 \pm \sqrt{1 - 2 \cdot \hat{q} + \frac{2 \cdot \hat{q}}{n}}}{n - 1}. \tag{11}$$

Taking into consideration that $\hat{p}$ is probability of the bit error and $0 \leq \hat{p} \leq 1$, one should take only the solution which includes the minus sign in place of $\pm$ in the above equation. However, in Fig. 6 one can see the second part of the solution which forms a characteristic square root curve. Each curve has a certain point that corresponds to the maximum value of $p$. This point $(q_0, p_0)$ can be determined under the condition that the expression under the square root operator sign should be zero, obtaining the following result:

$$(q_0, p_0) = \left( \frac{n}{2 \cdot (n-1)}, \frac{1}{n-1} \right). \tag{12}$$

Several initial values of $(q_0, p_0)$ pairs for increasing level of redundancy are presented in Table 1.

**Table 1.** Coordinates of $(q_0, p_0)$ for BER vs $q$ curves

|       | $n = 2$ | $n = 3$ | $n = 4$ | $n = 5$ | ... | $n \to \infty$ |
|-------|---------|---------|---------|---------|-----|-----------------|
| $q_0$ | 1       | 3/4     | 2/3     | 5/8     |     | 1/2             |
| $p_0$ | 1       | 1/2     | 1/3     | 1/4     |     | 0               |

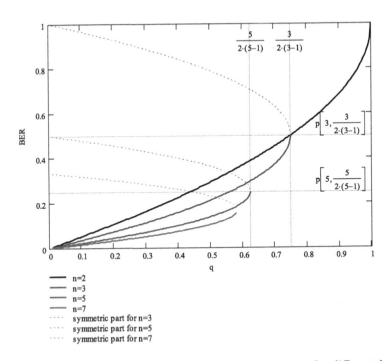

**Fig. 6.** BER based on the probability of incorrect bit detection for different levels of redundancy

# 3   Conclusions

Reliable communication does not require complex devices, protocols and standards, especially for wireless sensors that periodically sense the environment and transmits data to the sink node. Most of sensors can only be equipped with transmitter that reduces the power consumption as well as the cost of WSN but should fulfill the requirements for the network project parameters, i.e. maximum range, SNR, number of nodes, which implies demand for bandwidth, etc. Since each node can transmit at the same time, wireless signal can interfere which is manifested by errors in the sink node. Therefore, one should use some FEC technique that at least detects incorrect data. It is proposed to use the simplest method based on multiple transmissions of the same packet. Additionally, it is possible to correct wrong bits received during worse SNR conditions. It is not bandwidth efficient, but resistant to errors that occur in bursts, especially when interleaving is implemented. The only drawback is that detected errors could not be directly mapped to BER measure, which is the main indicator of data reliability. In this article, it has been proposed how to determine and calculate its value for different levels of redundancy using derived closed form expression. Since the form of the expression is quite simple, it could be easily implemented in any receiver.

# References

1. Akyildiz, I.F., Su, W., Sankarasubramaniam, Y., Cayirci, E.: Wireless sensor networks: a survey. Comput. Netw. **38**, 393–422 (2002)
2. Bandyopadhyay, S., Bhattacharyya, A.: Lightweight internet protocols for web enablement of sensors using constrained gateway devices. In: International Conference on Computing, Networking and Communications, pp. 334–340 (2013)
3. Begen, A., Roca, V.: Forward Error Correction (FEC) Framework. RFC 6363 (2011)
4. Ben-Tovim, E.: ITU G.hn - broadband home networking. In: Berger, L., Schwager, A., Pagani, P., Schneider, D. (eds.) MIMO Power Line Communications: Narrow and Broadband Standards, EMC, and Advanced Processing. Devices, Circuits, and Systems. CRC Press, Boca Raton (2014)
5. Chen, L.-J., Sun, T., Sanadidi, M.Y., Gerla, M.: Improving wireless link throughput via interleaved FEC. In: Ninth International Symposium on Computers and Communications (ISCC 2004), pp. 539–544 (2004)
6. Gondo, C.: Unified turbo/LDPC code decoder architecture for deep-space communications. IEEE Trans. Aerosp. Electron. Syst. **50**(4), 3115–3125 (2014)
7. Lake, D., Rayes, A., Morrow, M.: The internet of things. Internet Protoc. J. **15**(3), 10–19 (2012)
8. Singh, A., Chandran, H.: Low complexity FEC systems for satellite communication. Netw. Protoc. Algorithms **4**(1), 58–68 (2012)
9. Tanenbaum, A.S., Wetherall, D.J.: Computer Networks, 5th edn. Pearson, Upper Saddle River (2010)
10. Vacirca, F., De Vendictis, A., Baiocchi, A.: Optimal design of hybrid FEC/ARQ schemes for TCP over wireless links with Rayleigh fading. IEEE Trans. Mob. Comput. **5**(4), 289–302 (2006)

# Power Consumption Optimization in Datacenters Using PSO Tuning in Fuzzy Rule-Based Systems

Rocio Perez de Prado[1(⊠)], Jose Enrique Munoz-Exposito[1],
Sebastian Garcia-Galan[1], C. Mora Garcia[1], and Adam Marchewka[2]

[1] Universidad de Jaen, EPS Linares, CCTL Avenida de la Universidad, Jaen, Spain
`rperez@ujaen.es`
[2] Institute of Telecommunications and Computer Science,
UTP University of Science and Technology in Bydgoszcz, Bydgoszcz, Poland
`adimar@utp.edu.pl`

**Abstract.** This paper presents a strategy for reducing power consumption in a data centers in cloud computing. A more efficient use of resources using optimal scheduling of tasks is proposed. The scheduling strategy uses a fuzzy rule-based system (FRBS) with automatic learning for knowledge adquisition. The learning strategy is inspired on Particle Swarm Optimization algorithm and it allows the tuning of fuzzy sets of the FRBS without the need for obtaining new rules in a way that the initial rule base introduced by an expert is maintained through the whole performance of the scheduler.

## 1 Introduction

Cloud computing is an trending information technology nowadays which has become one of the most relevant large-scale distributed computing [9]. The most important advantage is the easy access to information and services from any-where, lower hardware costs and software, and diversity of devices for managing information. However, the main disadvantage is the associated power consumption. In order to provide all cloud services it is necessary that cloud suppliers have large data centers, that consume a lot of electricity (high power consumption). This consumption produces environmental pollution and in this way it is important to improve their performance. According to the studies by Jonathan G. Koomey [6,7], the increase of energy in data centers from 2005 to 2010 was 56 %, and have already doubled the energy consumption between 2000 and 2010, corresponding to 1.1 % and 1.5 % of total energy consumed. Over these years this rate has been increasing, and it is expected to be even greater according to storage and management data growth of the Internet of Things (IoT), which provides a total of 26 billion connected devices by 2020, according to tech consultancy Gartner. This means that the energy consumption in data management will be something to consider. The use of photovoltaic solar and wind energy will reduce pollution and $CO_2$. Renewables are a very important to reduce the impact of

© Springer International Publishing AG 2017
R.S. Choraś (ed.), *Image Processing and Communications Challenges 8*,
Advances in Intelligent Systems and Computing 525, DOI 10.1007/978-3-319-47274-4_32

data centers, but they are not the only solution. This paper presents a strategy for improving the scheduling of tasks, that reduces the power consumption on data centers.

This paper is organized as follows. Section 2 establishes some basic notation and background. Section 3 formally introduces the proposed genetic algorithm. In Sect. 4, simulations and results are shown and finally in Sect. 5 we conclude with final remarks and conclusions.

## 2 Background

In this section a short overview for cloud computing and cloud scheduling is presented.

### 2.1 Cloud Computing

Cloud computing is a current trend that involves the supply of computing resources as services. Major suppliers provide services to multiple customers remotely via Internet. Three types of services offered in the cloud can be differentiated:

- Infrastructure as a Service, IaaS.
- Platform as a Service, PaaS.
- Software as a Service, SaaS.

This paper focuses on the IaaS layer. It is based on the outsourcing of data processing machines and storage model. With this model a separation between the perceived infrastructure by users and the real systems where operations are performed is obtained. Iaas uses virtualization platforms, the goal of virtualization is to imitate the hardware implementation needed. Using these programs, we can create 'virtual machines' where operating systems and software required can be installed. Some applications to create virtual machines are VMware or Virtualbox.

Using virtualization presents some advantages:

- Security. If an installed application in a virtual machine is vulnerable, the attack will take place only in a restricted area, affecting the associated resources of that machine.
- Backup system. It can be programmed to back up the entire virtual machine. Also, a limited backup just enough to re-start the machine is possible.
- Robustness to hardware errors. If the physical system (server) hosting the application virtualization and virtual machines fail, another server and virtual machines copy can substitute them.
- Ease to manage migrations.
- Testing applications installed in the virtual machine without instability problems, since save points can be used within the machine and in this way, it is possible to undo the last performed installation.

## 2.2   Cloud Scheduling

In cloud computing optimization, the efficiency must take into account in two points of view:

- Companies must provide sufficient resources to the user, to have a fast and convenient experience. The company optimizes the execution time increasing the number of users served.
- On the other hand we must reduce power consumption. Data centers have high power consumption. Thus it will be making profitable the available resources and a efficiently use of them.

Since virtualization is a great method for saving resources, it will be the strategy used for power optimization. This will include allocation of virtual machines, and this mapping can be performed by heuristics schedulers or strategies based on artificial intelligent, for example fuzzy rules based systems.

Heuristic scheduling is a paradigm that solves the assignment of virtual machines using heuristic algorithms that evaluates fitness functions. In results section, five strategies are presented, to compare and test the proposed strategy.

Fuzzy logic systems address the imprecision of the input and output variables by defining fuzzy numbers and fuzzy sets that can be expressed in linguistic variables (e.g. small, medium and large). Fuzzy rule-based approach to modelling is based on verbally formulated rules overlapped throughout the parameter space. They use numerical interpolation to handle complex non-linear relationships. Many of existing systems need the rules to be formulated by an expert. However rules can be also generated automatically on the basis of numerical data. To obtain knowledge, there exist different strategies. Pittsburgh [12] and Michigan [2] are based on the use of evolutionary algorithms [3] where learning fuzzy rules is performed. Another strategy is KASIA (Knowledge Acquisition Approach with a Swarm Intelligence), and it performs learning of fuzzy rules, but is based on an a acquisition of knowledge algorithm using swarm intelligence (PSO).

## 3   Proposed Algorithm

This paper presents an scheduling algorithm based on tuning fuzzy sets on a FRBS [3]. Tuning uses a strategy based on PSO and it is named as PSO tuning (PSOT). PSOT is used for optimization in FRBS, it will focus on the optimization of better fuzzy sets.

The particle swarm optimization or PSO has demonstrated its utility in a wide range of research areas [1,4,11,13]. It has been experimentally shown that PSO achieve rapid convergence. Moreover, PSO algorithm is based on an a stochastic evolution of swarm intelligence and has proven to outperform other optimization algorithms such as genetic algorithms. One of the main advantages of these algorithms is related to its simple implementation and reduced number of binding parameters. PSO not need genetic operators (i.e. crossing or mutation) because the particles are updated themselves considering its internal speeds.

Furthermore, the particles have memory and consider exchanges information, which in particular reduces the number of communications between particles. Moreover, it can exercise greater control over the convergence of the particles compared to genetic algorithms, which considerably decreases the number of required assessments.

In canonical PSO, each individual is known as a "particle" and it moves within a multidimensional space representing the search space. The system is initialized with a set of $M$ particles randomly distributed in the n-dimensional search space $\mathbb{R}^N$. Furthermore, a real function $f$ is defined in such space constituting the fitness function, $f : \mathbb{R}^N \to \mathbb{R}$. Each particle position is modified through iterations with the aim of finding the optimum position, where an optimum value for the fitness function or optimum state is achieved. This modification is done on the basis of three components leading the whole process: an inertial component, a self-recognition component or an inner tendency to return to its best position, and a social component representing the swarm particle leaning to move toward the best position found by its neighbors. Thus, at every iteration $(t+1)$, position $x$ and velocity $v$ of particle $i$ are updated by the following expression:

$$v_i^{(t+1)} = \omega v_i^t + d_1 r_1 (Pb_i^t - x_i^t) + d_2 r_2 (Gb^t - x_i^t) \tag{1}$$

$$x_i^{(t+1)} = x_i^t + v_i^{(t+1)} \tag{2}$$

where $d_1$, $d_2$ are constant weight factors, $Pb_i$ represents the best position achieved by particle $i$, $Gb_i$ indicates the best position found by the neighbors of particle $i$, $r_1$, $r_2$ are random factors in the [0,1] interval, and $\omega$ denotes inertia weight. Moreover, to ensure the algorithm convergence, $v$ values are constrained to the interval $[-v_{max}, v_{max}]$. Furthermore, another key factor for convergence is to set efficiently the inertia weight $\omega$. A high value for $\omega$ favors the global search, whereas a low value favors local search. In other words, $\omega$ can be configured to balance both types of search and, thus, to reduce the number of required operations to reach the optimum value. Hence, $\omega$ is usually associated to a decreasing value through iterations in a way that local searches are intensified once the whole space has been explored. Typically, $\omega$ is initialized with a value that is close to the unity.

The proposed PSOT algorithms considers a particle as a fuzzy set (Fig. 1). A fuzzy set is characterized according to two parameters: mean and standard deviation. Therefore a PSOT particle has the structure shown in Eq. 3.

$$P_i = \begin{pmatrix} td_{1,1}^i & mean_{1,2}^i \\ ... & ... \\ std_{m,1}^i & mean_{m,2}^i \end{pmatrix} \tag{3}$$

where std is the standard deviation of the fuzzy set, mean is the mean of the set, i is the number of particle and n is the number of parameters.

The particles are randomly generated, within a limited workspace. Its dimension is fixed and it is linked to the number of variables and the fuzzy set of each

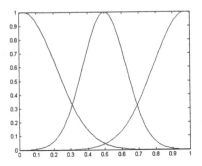

**Fig. 1.** Fuzzy set

variable. The particles change their position according to the velocity in Eq. 1, considering that the velocity of each particle is determined by Eq. 4

$$V_i = \begin{pmatrix} V_{1,1}^i & V_{1,2}^i \\ \dots & \dots \\ V_{m,1}^i & V_{m,2}^i \end{pmatrix} \tag{4}$$

The variation of each average particle changes the position of fuzzy sets, while the variation of the standard deviation changes the shape (Fig. 2). All particles and velocities are initialized within the workspace, but some restrictions are needed, due to the value of updated particles can exceed the results domain. The algorithm selects a $\delta$ value such $|P_i| < \frac{P_i * \delta}{4}$. The process continues until the stop condition is reached.

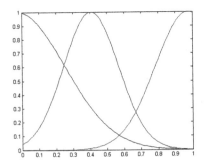

**Fig. 2.** Tuning fuzzy sets using PSOT

The proposed tuning scheduler algorithm based on PSO is summarized below.

**PSOT algorithm**

    1. Swarm: Num_particles,
                Num_iter, Init_rate $(r_0)$, Inertial_weight $\omega$, Factors $c_1$ and $c_2$.
    2. Random setting of Swarm position.
    3. Random setting of velocity.
    4. Velocity Constraints.
    5. Initialize Gbest$(P^*)$/Pbest$(P^\#)$
**Do**
    **Do**
        1. Update position. Eq. 2.
        2. Constraints RB-Swarm position.
$$|P_i| < \frac{P_i * \delta}{4}$$
        3. Evaluate-Fitness. Evaluation system.
      Particles ++
    **While**(Num_particles)
      Update Gbest $(P^*)$.
    **Do**
        1. Update Pbest $(P^\#)$.
        2. Update velocity. Eq. 1.
        3. Velocity Constraints.
      Particles ++
    **While**(Num_particles)
    iter++
**While**(Num_iter)
**Return** solution: Gbest $(P^*)$

# 4  Simulation and Results

Several tests have been conducted to evaluate the proposed scheduling scheme. Concretely, the scheduler is tested through simulations scenarios and traces from real world. In our tests, the environment is based on Real CloudSim Java simulator [8].

Moreover, in order to test the efficiency of proposed system, the results of simulation are compared with other literature classic scheduling algorithms: Random, Round Robin (RR), MinMin, Maxmin and Sufferage [5]. In the Round Robin algorithm, each virtual machine is allocated in a different host, making a cycle. Hosts that cannot allocate virtual machines are skipped. In the Best Resource Selection algorithm, the host with the highest MIPS in use divide MIPS ratio is allocated. MinMin allocates virtual machine in the host with minimum power after allocation. MaxMin allocates virtual machine in the host with maximum power after allocation. The Sufferage heuristic is based on the idea that better mappings can be generated by assigning a virtual machine to a host that would suffer most in terms of expected power consumption if that particular virtual machine is not assigned to it. The results are compared with other FRBS schedulers: KASIA [10] and Pittsburgh [12].

The simulation environment consisting of 100 nodes with 250 virtual machines and 250 tasks to serve the system. Nodes, virtual machines and tasks are heterogeneous: all have different characteristics. To perform a reliable study have been conducted 30 experiments with each scheduler.

Five variables related to power consumption have been selected for scheduling. The variables are:

- MIPS. Microprocessor without Interlocked Pipeline Stages.
- Pow. Power consumed in a node
- CPUutil. CPU utilization.
- Powalloc. Power consumed after allocation of a virtual machine.
- Utilalloc. CPU utilization after allocation of a virtual machine.

The FRBS has 5 inputs variables with 3 membership functions and 1 output variable with 5 membership functions. The rules base used is:

1. IF (mips IS low) OR (pow IS low) OR (utilalloc IS medium) THEN out IS NOT low
2. IF (mips IS low) OR (pow IS NOT low) OR (cputil IS hight) OR (powalloc IS hight) THEN out IS low
3. IF (mips IS medium) OR (pow IS medium) OR (utilalloc IS low) OR (powalloc IS hight) THEN out IS low
4. IF (mips IS NOT low) OR (pow IS hight) OR (cputil IS medium) (utilalloc IS NOT low) THEN out IS low
5. IF (mips IS medium) OR (cputil IS hight) OR (utilalloc IS NOT low) OR (powalloc IS hight) THEN out IS medium

Table 1 presents an energy power consumption comparison between these algorithms, compared to the algorithm proposed in this work. Results suggest that PSOT performs better than heuristic. It is shown that PSOT scheduler achieves the best performance in terms of power consumption in comparison to heuristic algorithms and similar to KASIA and Pittsburht results.

**Table 1.** Power consumption results

| Power (kW) | | | | | | | |
|---|---|---|---|---|---|---|---|
| Heuristic scheduling | | | | | FRBS | | |
| Random | Round Robin | MinMin | MaxMin | Sufferage | Pittsburgh | KASIA | PSOT |
| 6.568 | 6.990 | 6.531 | 6.569 | 6.631 | 6.293 | 6.330 | 6.313 |

Figure 3 shows the convergence for proposed algorithm. A summary of power consumption is presented in Fig. 4. It can be observed that exists a significant difference between evolutionary and heuristics strategies.

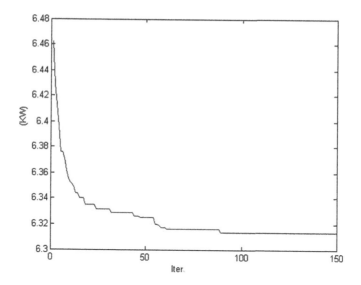

**Fig. 3.** PSOT convergence

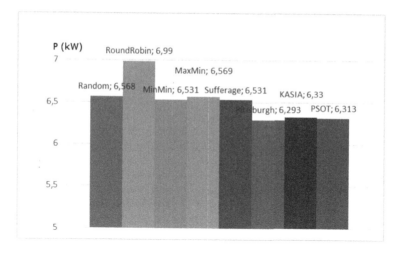

**Fig. 4.** Reglas del sistema FRBS

## 5   Conclusions

This paper proposes the development of a system based on FRBS tuning sched-uler to improve the allocation of virtual machines and tasks in IaaS data centers. Its use reduces the power consumption generated by this infrastructure.

Large data centers make use of a lot of electricity due to the required high power consumption for their performance. Since they are a source of air pollu-tion and other environmental problems and high companies costs, it is relevant to increase their efficiency in electricity management. In this work a scheduling algorithm for cloud computing has been presented. It uses a FRBS system for scheduling in homogeneus and heterogeneus datacenters and reduces power consumption. Results indicate that a FRBS tuning scheduler achieves better performance in terms of power consumption in comparison to classic scheduling algorithms: it reduces consumption respect to the best of its competitors in medium and large datacenters, respectively, and thus, it arises as a competitive green scheduling strategy for cloud computing.

**Acknowledgments.** This work was financially supported by Research Projects TEC2015-67387-C4-2 and TEC2012-38142- C04-03.

# References

1. Araujo, E., dos Santos Coelho, L.: Particle swarm approaches using lozi map chaotic sequences to fuzzy modelling of an experimental thermal-vacuum system. Appl. Soft Comput. **8**(4), 1354–1364 (2008)
2. Booker, L.B., Goldberg, D.E., Holland, J.H.: Classifier systems and genetic algorithms. Artif. Intell. **40**(1), 235–282 (1989)
3. Cordón, O.: Genetic Fuzzy Systems: Evolutionary Tuning and Learning of Fuzzy Knowledge Bases, vol. 19. World Scientific, Singapore (2001)
4. Hussein, T., Elshafei, A., Bahgat, A.: Comparison between multi-band and self-tuned fuzzy power system stabilizers. In: 2008 16th Mediterranean Conference on Control and Automation, pp. 374–379. IEEE (2008)
5. Kim, T., Adeli, H., Cho, H., Gervasi, O., Yau, S.S., Kang, B.-H., Villalba, J.G. (eds.): GDC 2011. CCIS, vol. 261. Springer, Heidelberg (2011)
6. Koomey, J.: Growth in data center electricity use 2005 to 2010. A report by Analytical Press, completed at the request of The New York Times 9 (2011)
7. Koomey, J.G.: Worldwide electricity used in data centers. Environ. Res. Lett. **3**(3), 034008 (2008)
8. Lab, M.C.: CloudSim (2016). http://www.cloudbus.org/cloudsim/
9. Mell, P., Grance, T.: The nist definition of cloud computing (2011)
10. Prado, R., Exposito, J.M., Yuste, A., et al.: Knowledge acquisition in fuzzy-rule-based systems with particle-swarm optimization. IEEE Trans. Fuzzy Syst. **18**(6), 1083–1097 (2010)
11. dos Santos Coelho, L., Herrera, B.M.: Fuzzy identification based on a chaotic particle swarm optimization approach applied to a nonlinear yo-yo motion system. IEEE Trans. Ind. Electron. **54**(6), 3234–3245 (2007)
12. Smith, S.F.: A learning system based on genetic adaptive algorithms (1980)
13. Venayagamoorthy, G.K., Grant, L.L., Doctor, S.: Collective robotic search using hybrid techniques: fuzzy logic and swarm intelligence inspired by nature. Eng. Appl. Artif. Intell. **22**(3), 431–441 (2009)

# Considering Service Name Indication for Multi-tenancy Routing in Cloud Environments

Sebastian Łaskawiec$^{(\boxtimes)}$ and Michał Choraś

Institute of Telecommunications,
University of Science and Technology in Bydgoszcz,
Bydgoszcz, Poland
sebastian.laskawiec@gmail.com, chorasm@utp.edu.pl

**Abstract.** In recent years Cloud deployment gained popularity mainly because of reducing IT operational costs. Cloud vendors and middleware companies work hard on hosting their services for many customers at the same time. This requires solving two fundamental problems: how to isolate customer data one from an- other and how to securely access it. This article shows how to use TLS/SNI mechanism to create a multi-tenant router which will be used in Infinispan project.

**Keywords:** Cloud · PaaS · SaaS · Multi-tenant applications · Hosted software

## 1 Introduction

The simplest way to host services in Cloud environment is to use Infrastructure as a Service (IaaS). This solution has been used starting from 2003 when Xen hypervisor was invented. Hosting multiple guest operation systems requires a lot of physical resources (memory, disk I/O and CPU). In 2008 Linux Containers project addressed this problem introducing lightweight system virtualization based on Linux CGroups and Chroot. Seven years later Docker combined LXC project concepts with layering and packaging and became a new standard for virtual machines. Projects like Docker and Kubernetes are used as a foundation of modern cloud environment such as Google Compute Engine or OpenShift by Red Hat [1].

With on-demand application scaling, a new type of services can be sold by Cloud vendors - Platform as a Service. This allows vendors to utilize resources better and maintain a single cluster instead of multiple installations. This setup performs very well for stateless applications where customer passes data into the service and gather results (for example a service for sending emails). However serving data stores is much more complicated and requires separating data per each tenant. The second concern is to maintain security context for accessing the data so that each tenant could access only his own partition.

R.S. Choraś (ed.), *Image Processing and Communications Challenges 8*,
Advances in Intelligent Systems and Computing 525, DOI 10.1007/978-3-319-47274-4_33

Unfortunately there is no generic solution for the first part of the problem (isolating data) and each software vendor needs to come up with his own ideas. Database vendors often use database schema approach (schema per tenant) whereas data grid vendors can separate data store clusters from each other. The second part (security context) is much more interesting. A lot of software deployed in the Cloud uses REST as a main interface. However HTTP messages contain a lot of overhead and gaining maximum performance requires using transport protocols and sockets directly (UDP and TCP). The most popular security protocol that can be used with this setup is SSL/TLS. Additionally a TLS extension called Service Name Indication (SNI) allows transmitting client host name and choosing proper key store on server side. The same mechanism can be reused for choosing proper data partition simplifying data isolation aspect. This article explains the idea of separating applications in paragraph two. In the third paragraph the router concept is explained in details and high performance application requirements are discussed in paragraph four. Paragraphs five and six explain how SNI works and how to use it in multi-tenant environment. The implementation and evaluation results were gathered in paragraphs seven and eight followed by the conclusion.

The value added of this paper is an evaluation of the idea of using transport layer security (SSL/TLS) for a multi-tenant routing with implementation and evaluation of the results.

## 2 Separating Applications in Cloud Based Environment

Separating hundreds of applications within a Cloud is complicated and challenging. Cloud vendor needs to control whether a hosted application can or can not call each other's endpoints and manage security context in which such communication can occurs. Modern Cloud management solutions often use Docker (or Linux Containers in general [2]) as a containers for user applications. This approach has a lot of advantages against traditional Virtual Machines, which is performance [3] and portability (LXC is supported out of the box by Linux OS).

Running multiple guest applications (or multiple guest operation systems in general) requires separating them one from another. One of the simplest ways how to deal with it is to divide applications by tenants (Fig. 1). Applications within one tenant should be able to should not.

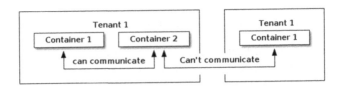

**Fig. 1.** Multi-tenant communication

Tenants can be coarse grained like companies or fine grained like projects or departments within a company.

The separation is often done in layer 2 and 3 of the OSI model and over the years the industry came up with different standards to achieve that [4] but the high level concept remains the same. Each VM (or container) is attached to a virtual NIC and packets are routed through a virtual switch. The implementation needs also to handle layer 3 and moving VM from one compute node to another. Some of the Cloud vendors mention what technology is used in their products, for example OpenShift by Red Hat uses Open vSwitches [5] and Google Cloud Platform developed their own project called Andromeda [6].

## 3 Multi-tenant Applications

Separating applications from other tenants is not always the case. Sometimes the Cloud vendors (or even tenants) want to share a single application for multiple clients. The most common example is sharing a database (Fig. 2). This model has also been used by many organizations on premise - a single database installation serves data for multiple projects.

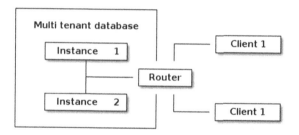

**Fig. 2.** Multi-tenant application

One of the biggest challenges of supporting multi tenancy is how to separate user data from each other. Database industry came up with a concept of database schema [7]. It allows to host multiple data containers in a single database instance. This approach requires a clear separation between data hosting layer and transport in the application. This approach might not always be valid for high performance, data intensive application where each tenant might have different transport requirements. In such case there needs to be a way to define transport related settings per tenant.

## 4 Using SNI Server Name for Identifying Client

The Transport Layer Security (TLS) is responsible for establishing (or resuming) secure session between a server and a client. In order to decrypt the data both server and client need to follow a handshake procedure which involves the following:

1. Negotiating cipher suite
2. Authentication
3. Exchanging keys

When designing services for Cloud which will be consumed in PaaS fashion, the server needs to use different encryption keys for each client while its IP address remains the same. Similar problem occurs when hosting multiple virtual web servers on a single IP address and using the host header for multiplexing. For TCP or UDP based transports the host header is encrypted and therefore the server can not read it before finishing TLS handshake. In order to accomplish the handshake the server needs to choose proper certificate based on clients hostname (Service Name Indication) [8,9]. A simplified flow diagram might be found on Fig. 3.

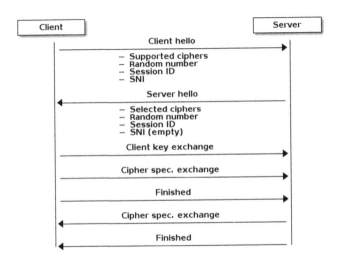

**Fig. 3.** TLS handshake with SNI

An obtained SNI host name can be easily used for both authentication and choosing proper tenant for hosted service. The SNI approach for routing has an additional advantage - when the client uses wrong credentials, it is being reported as handshake errors (which makes a potential attack vector smaller as the attacker does not know if given tenant exists or the credentials are incorrect).

## 5    SNI Based Multi-tenant Router Implementation

In a typical Cloud environment, hosted services are separated from the outside network and additional routing component is needed to forward requests to inner servers from the clients (Fig. 2). Services deployed in the Cloud need also to isolate tenants from each other. This problem is typical for all data containers including databases, in memory stores (data grids), caches etc. Some applications allow hosting multi data containers within a single process. Infinispan HotRod Server is one of such applications however each data container (an 'EmbeddedCacheManager' to be accurate) uses its own TCP Socket and binds to a different port (Fig. 4).

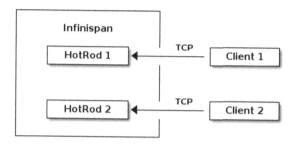

**Fig. 4.** Multiple data containers inside single hot rod process

The multi-tenant router separates HotRod servers from TCP Servers and starts a single frontend for all inner services (Fig. 5). This approach has smaller resource consumption (opposed to starting a frontend server per each HotRod) and completely decouples forwarding requests from data container operations.

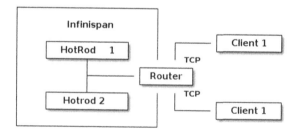

**Fig. 5.** Multiple data containers and a router

Each TCP Server uses Netty framework which is based on handlers architecture. All handlers are gathered together in a concept called a Pipeline where each handler is responsible for a piece of work (for example encrypting data using SSL or decoding data into user defined structures) [10]. This kind of pattern is very common when designing network protocols and many popular libraries use it (JGroups framework for example).

The router uses concept called Routes for forwarding datagrams from input to proper HotRod instance. Each route consists of 'RouteSource' and 'RouteDesitnation'. A 'RouteSource' is responsible for recognizing used network protocol and gathering all necessary information to distinguish clients - in case of SNI based 'RouteSource' this would be also SNI Host Name. A 'RouteDestination' is responsible for forwarding data to proper component - in case of HotRod server it takes all its handlers and adds them to the pipeline (Fig. 5).

*Routing function implementation pseudo-code*

```
Input:  Inbound TLS/SNI connection ic
    SNI Handler sh
    Routing table rt
    Connection pipline p
Output: RouteDestination rd

if(sh.handshakeAccepted(ic))
  shn = sh.getHostname()
  Collection<Route> r = rt.getRoutes()
  // Use lazy evaluation for routes
  Route rd = r
      .filter(ri -> ri.source() is SNISource)
      .filter(ri -> ri.sniHostName() == shn)
      .filter(ri -> ri.desitnation() is HotRodDestination)
      .getFirst()
  if(rd == null)
    throw Exception("No Route")
  pipeline.removeHandler(sh)
  pipeline.addHandlers(rd.sestination().handlers())
else
  throw Exception(``No TLS/SNI Connection'')
```

The algorithm is very straightforward and its complexity is proportional to the number of routes in the routing table. Using a Big-O notation it might be written as:

$$T(n) = O(n). \tag{1}$$

An optimized implementation has been proposed as a multi tenancy implementation for Infinispan project [11].

## 6   Prototype Evaluation

The basic implementation allows connecting TLS/SNI based connections to a HotRod server. A RouteSource and a RouteDestination interfaces are generic enough to connect all various types of services. However this flexibility comes with a cost of performance. In order to determine how much overhead it adds, a JMH [12] test was performed and the results were gathered in tables below [13]. The test server configuration was the following: Linux OS (Fedora F23), Intel®Core i7-4900MQ CPU @ 2.80 GHz, 16 GB RAM (Tables 1 and 2).

**Table 1.** Initializing connections

| Test description | Calculation method | Number of iterations | Test result | Units |
|---|---|---|---|---|
| Single server without SNI | Average | 31 | 3.046 | ms/op |
| Single server with TLS/SNI | Average | 31 | 12.725 | ms/op |
| 2 servers and SNI router | Average | 31 | 13.471 | ms/op |

**Table 2.** Performing 10 000 Cache PUT operations (op in this context is equal to 10 000 put operations)

| Test description | Calculation method | Number of iterations | Test result | Units |
|---|---|---|---|---|
| Single server without SNI | Average | 31 | 430.075 | ms/op |
| Single server with TLS/SNI | Average | 31 | 847.909 | ms/op |
| 2 servers and SNI router | Average | 31 | 1022.655 | ms/op |

# 7   Conclusions

New application development and deployment trends like virtualization, Docker containers or hosting applications as a service gain popularity very quickly [1]. When considering data store applications like databases or data grids, hosting a single service for multiple tenants is one of the best ways to lower operational costs and increase user experience. Multi tenancy is not an easy thing to implement though and requires additional care for separating data. Since hosted services are deployed in the Cloud and many configurations require advanced routing (like Hybrid Clouds), it is a good custom to encrypt all connections. The TLS/SNI router implementation turns out to be helpful when addressing both those concerns at the same time. TLS ensures the connection is encrypted and SNI allows to recognize tenant and helps this way choosing the right partition to access. There are some scenarios where additional routing step is required when TLS is not used (in that case a custom protocol can be used or when considering HTTP transport, some additional headers might be added to the HTTP datagram). How- ever a generic router implementation should be flexible enough to support this scenarios as well (additional RouteSource implementation are required in that case). Using TLS/SNI for routing adds some additional overhead for communication (more than 1 milliseconds for initializing connection and 175 milliseconds for performing 10 000 operations). Additional delay for initializing connection is expected and usually do no harm but adding 17 microseconds per each operation might be a bit too high for low latency environments (and data grids should be considered as such). Route filtering might seem like a good candidate for optimization, but it can speed up only initializing connections (after the handshake, the routing handler removes itself from the pipeline). Exposing settings for tuning connection pools as well as using EPoll are much better candidates for optimization. Servers like Infinispan HotRod are designed carefully for optimal performance and the router should expose as many configuration parameters as possible to allow fine tuning for achieving best possible results in each scenario.

# References

1. Łaskawiec, S.: The evolution of Java based software architectures. J. Cloud Comput. Res. **2**(1), 1–17 (2016). Columbia International Publishing, Columbia
2. The Linux Containers project. https://linuxcontainers.org
3. Felter, W., Ferreira, A., Rajamony, R., Rubio, J.: An updated performance comparison of virtual machines and Linux containers. In: International Symposium on Performance Analysis of Systems and Software, USA, pp. 171–172 (2015)
4. Network Virtualization and Software Defined Networking for Cloud Computing: A Survey. http://www.cse.wustl.edu/~jain/papers/ftp/net_virt.pdf
5. OpenShift Product Documentation. https://docs.openshift.com/enterprise/3.0/architecture/additional_concepts/sdn.html
6. Enter the Andromeda zone - Google Cloud Platform's latest networking stack. https://cloudplatform.googleblog.com/2014/04/enter-andromeda-zone-google-cloud-platforms-latest-networking-stack.html
7. Laszewski, T., Nauduri, P.: Migrating to the Cloud: Oracle Legacy Client/Server Applications. Syngress, Waltham (2011). ISBN 9781597496476
8. Shbair, W.M., Thibault, C., Goichot, A., Chrisment, I.: Efficiently bypassing SNI-based HTTPS filtering, efficiently bypassing SNI-based HTTPS filtering. In: Shbair, W.M., Cholez, T., Goichot, A., Chrisment, I. (eds.) IFIP/IEEE International Symposium on integrated Network Management (IM 2015), Canada, pp. 990–995 (2015)
9. TLS Handshake Protocol description, Microsoft. https://msdn.microsoft.com/pl-pl/library/windows/desktop/aa380513(v=vs.85).aspx
10. Mauer, N., Wolfthal, M.A.: Netty in Action. Manning Publications, Greenwich (2014). ISBN 9781617291470
11. Multi tenancy implementation proposal. https://github.com/infinispan/infinispan/pull/4348
12. JMH microbenchmark tool for JVM based languages. http://openjdk.java.net/projects/codetools/jmh
13. Benchmark results. https://gist.github.com/slaskawi/51f76b0658b9ee0c9351bd172 24b1ba2

# Erratum to: A Flexible Software Architecture for a Network of Heterogeneous Smart Cameras

Dominik Pieczyński, Marek Kraft[✉], and Michał Fularz

Institute of Control and Information Engineering,
Poznań University of Technology, Piotrowo 3a, 60-965 Poznań, Poland
marek.kraft@put.poznan.pl

**Erratum to:**
**Chapter 11 in: R.S. Choraś (ed.)**
**Image Processing and Communications Challenges 8**
**DOI: 10.1007/978-3-319-47274-4_11**

In the original version, the name of the first author was incorrect. Instead of "Dominik Pieczński" it should read as "Dominik Pieczyński". The original chapter was corrected.

---

The updated original online version for this chapter can be found at
DOI: 10.1007/978-3-319-47274-4_11

© Springer International Publishing AG 2017
R.S. Choraś (ed.), *Image Processing and Communications Challenges 8,*
Advances in Intelligent Systems and Computing 525, DOI 10.1007/978-3-319-47274-4_34

# Author Index

© Springer International Publishing AG 2017
R.S. Choraś (ed.), *Image Processing and Communications Challenges 8*,
Advances in Intelligent Systems and Computing 525, DOI 10.1007/978-3-319-47274-4

Printed in the United States
By Bookmasters